HARTMAN ON SKEET

HARTMAN ON SKEET

BARNEY HARTMAN

Stackpole Books

HARTMAN ON SKEET

Published by
STACKPOLE BOOKS
Cameron and Kelker Streets
Harrisburg, Pa. 17105

First published 1967 by McClelland and Stewart Limited, 25 Hollinger Road, Toronto and printed in Canada. Published 1973 by Stackpole Books and reprinted in the U.S.A. by special arrangement with McClelland and Stewart Limited.

Library of Congress Cataloging in Publication Data

Hartman, Barney.
Hartman on skeet.

1. Trap-shooting. I. Title.
GV1181.H3 1973 799.3'13 73-13806
ISBN 0-8117-0791-1

Contents

1. What is skeet? *8*
2. Gun fit *11*
3. Gun knowledge *27*
4. Guns — speaking personally *34*
5. To load a shot-gun *41*
6. Shooting equipment *42*
7. Good form *50*
8. Bedroom skeet *59*
9. Shooting station by station (singles) *60*
10. Shooting doubles *92*
11. What makes a good shooter? *98*
12. Acuity *101*
13. Stance for doubles at stations 3, 4, 5 *104*
14. How to call for a target *110*
15. How to start a new shooter *112*
16. What to look for when your scores drop *114*
17. Etiquette *116*
18. Tournament shooting *118*
19. Watching the targets *120*
20. Gun safety *122*
21. International skeet *125*
22. Flinching *132*
23. Targets *134*

Appendix *135*

Additional Comments *137*

Acknowledgement *140*

NOTE

For the sake of simplicity all instructions assume that the shooter is right-handed. Unless stated otherwise, procedures for left-handed shooters are simply reversed.

HARTMAN ON SKEET

1.

What is skeet?

A skeet field consists of two trap houses facing each other forty-two yards apart. Seven shooting stations are located around the edge of a semi-circle with an eighth midway between the two trap houses. Saucer shaped "clay" targets $4\frac{1}{4}$" in diameter and an inch high are propelled from either house along fixed lines of flight by a mechanical trap. They cross over a point six yards outside of station eight commonly called the **Crossing Point** *(refer to photos 1 & 2).*

In still air these targets must travel a distance of at least fifty-five yards to the **Target Distance Marker,** and pass over the crossing point at a height approximately fifteen feet above the level of the field. The house on the left of the field is the high house, commonly called the **Pull House,** and the targets emerge through an opening ten feet above ground level. From the low house, commonly called the **Mark House,** targets emerge through an opening only $3\frac{1}{2}$ feet above ground level. Twenty-five shots at twenty-five targets make a regulation round of skeet.

A squad consists of five shooters and each shooter in turn shoots at a target thrown from the high house and then one from the low house. Starting at station one directly in front of the high house the squad then proceeds around to stations two, three, four, five, six and seven. From these stations targets must be broken before they cross the boundary marker located at right angles out from the trap houses. The squad then moves to station eight where each shooter in turn shoots at a target from the high house and then a target from the low house. While shooting at this station, the targets must be broken before they pass over the crossing point. On returning to station one each shooter in turn then loads two shells to shoot doubles. Here the targets are released simultaneously and the shooter shoots at the high house outgoer first and then at the low house incomer. This is repeated at station two. The shooters then proceed to stations six and seven where they shoot at the low house outgoer and at the high house incomer. This accounts for twenty-four out of twenty-five shells and if the shooter has not missed a target he may then take either a high or low house target from whichever station he likes. Most shooters shoot either a high or low house target from

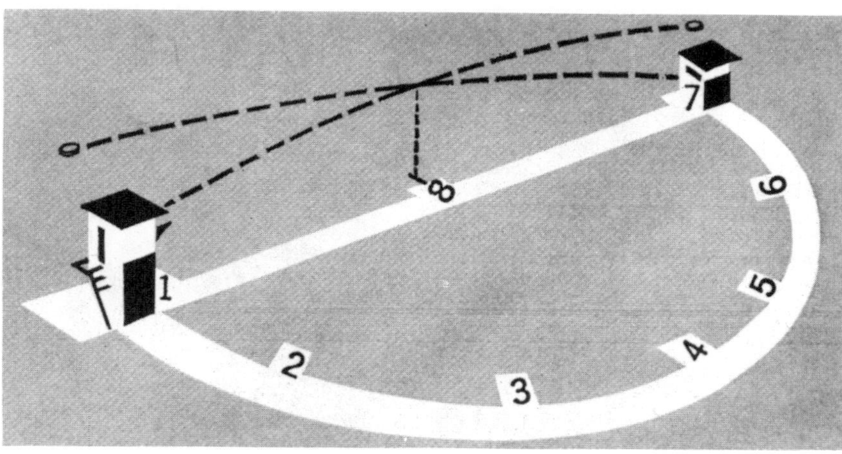

1. Simulated targets in flight

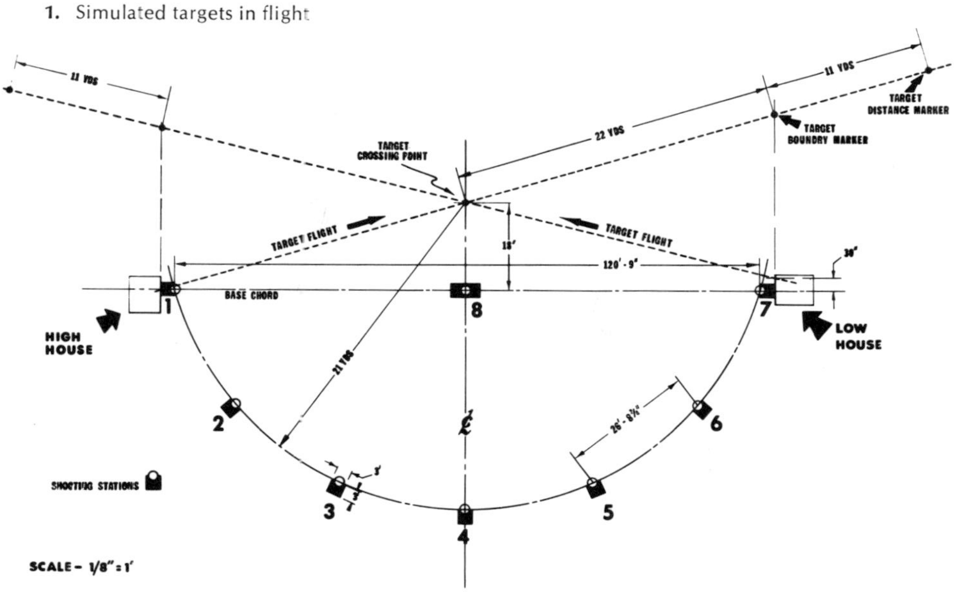

2. Skeet field layout

 station seven. Should a shooter miss a target in the round the optional shot is taken of the first target missed.

According to the rules a target is supposed to travel a minimum distance of fifty-five yards in still air and fifteen feet above field level at the target crossing point. One of the difficulties at most skeet fields is getting targets correctly thrown. Management and shooters fail to have proper regulation targets thrown; this occurs both during practice and at tournaments. In the winter months trap springs are adversely affected by the cold and unless they are tightened, the targets will fall far short of fifty-five yards. It is strongly recommended that this important responsibility of setting the targets correctly be given to knowledgeable shooters to ensure traps are properly maintained and set. With changes in wind direction and velocity further small changes are required in the traps to maintain the target on its flight-path. Remember, for distance, one of the stipulations is that the targets be set in still air. Targets can be thrown a full sixty-five yards and still be legal. If you set the traps to throw targets two to three yards past the fifty-five yard marker you can rest assured they will at least make the fifty-five yards even if thrown into a head wind of ten to fifteen m.p.h. Small electric heaters can be used on automatic traps – if they are turned on a half hour before shooting, targets will travel the regulation distance even in sub-zero temperature. When heaters are used it is not necessary to re-adjust the spring tension for winter shooting.

2.

Gun fit

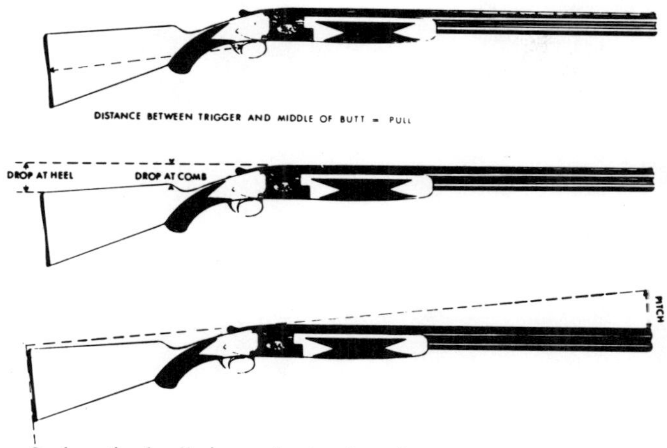

3. Length of pull, drop at heel and comb, and pitch

You've probably read a lot on how a gun should fit, and you may even have been advised to go to a good gun maker to be measured for a stock. Don't – it's just a waste of money when you're starting out. What's right for you at first won't fit after a year or so's shooting. Consulting a gun fitter is only worthwhile after you've had considerable shooting experience. For most skeet guns special stock fitting is out of the question. However, those who do wish to have special stocks made will need a "Try Gun" and a gun fitter. A try gun is nothing more or less than an ordinary gun fitted with a stock which can be adjusted laterally (cast), vertically (drop at comb and heel), lengthwise (as the name implies) and for pitch (muzzle down). To be measured for a gun is similar to being measured for a suit of clothes. The gun is required to be mounted several times and, after careful observation, the gun fitter will make several adjustments to the stock. Once he is satisfied, he will measure the stock and these measurements will be given to the stock maker who will turn out the finished product.

Some shooters blame the gun when their scores drop; they begin hacking away at their stocks, changing the drop or pitch, boring out the butt or adding weight to the fore-end. Some probably do not know what they are doing, but there are others who are mechanically inclined and by experimenting they learn a lot about the intricacies of proper gun fit. If you don't experiment you will never find out. Your improvement will be slow and your scores will never be consistent. The factors that determine whether or not a shot-gun fits properly are all found in the stock. They are: (1) length of pull, (2) drop at heel, (3) drop a comb, (4) pitch down and (5) cast (*refer to photo 3*).

"Length of Pull." The length of pull is the measured distance, usually in inches, from the centre of the butt or recoil pad to the front of the trigger. It is determined simply by the length of the stock. This measurement can be altered merely by cutting a piece off the butt or adding on a recoil pad. Skeet stocks usually come with 14" pull on automatics and pump guns, and 14¼" on some over and unders. Trap stocks on the other hand usually measure 14⅜".

The length of stock can only be found after a careful study on how the

4.

5.

shooter positions his gun on his shoulder, and how he uses that shoulder in conjunction with head position while under actual shooting conditions. Have the shooter place the butt of the gun on his shoulder; it should fit in the crotch of the shoulder and not out toward the arm. Make sure that the heel of the butt is high enough and not away down on the chest. The right hand should grasp the pistol-grip and the left hand should be properly positioned on the fore-end. The comb of the stock should nestle against the face and beneath the right eye. If the shooter is right-handed he should be looking down the rib of the barrel or just above it. To do this, the head should be slightly forward. Now, by standing to the side of the shooter, measure the distance from the tips of the fingers of the right hand (preferably the little finger) to the tip of the nose. This distance should be not less than half an inch and not more than 1½ inches *(refer to photos 4, 5, 6, 7 & 8)*. Don't take just one measurement, have the shooter assume the proper shooting position several times and if possible observe him under actual shooting conditions before making final measurements. Remember different types of shooting clothes, thickness of shoulder pads and heavier clothing worn in winter will affect these measurements. You may sometimes see a shooter come in with a bruised nose or cheek because his thumb or fingers have been in close proximity. Improper gun-stock length which leads to these injuries will also often precipitate flinching. A long stock, on the other hand, will destroy good

6.

7.

4. Gun stock too short; note nose and finger interference

5. Minimum length stock; approximately ½″ clearance measured by one finger

6. Maximum length stock; approximately 1½″ clearance measured by two fingers

7. Proper length of stock, shown without finger measurements

8. Stock too long; distance is 2″-2½″ from finger tips to nose

8.

 balance by throwing too much of the weight of the gun forward, thus creating stress in the forward arm. This makes the gun clumsy and uncomfortable to handle.

Usually a gun stock comes equipped with either a hard composition butt plate or a rubber recoil pad. To lengthen the stock, merely slip a few washers on the two screws between the stock and the butt plate *(refer to photo 9)*. This is only temporary and the stock may be lengthened up to 1/4 inch by this method. If this is found to be satisfactory a permanent fix can be obtained by fitting a 1/4 inch piece of Perspex in lieu of the washers. If more length is required thicker Perspex or a rubber recoil pad may be added if not already fitted. To shorten a stock 1/4 inch, simply remove the butt plate. Add or remove only small amounts at a time until the correct fit is obtained.

You may have seen shooters measuring themselves to see whether a stock is the correct length. They do this by placing the butt in the hollow of their arm and holding the pistol-grip with their right hand. The trigger-finger is placed on the trigger and if the stock touches the upper arm they consider the stock is correct. This is a poor substitute and should not be considered an accurate measurement.

"Drop at Heel and Comb." The next two measurements usually taken are drop at comb and heel. To be more exact, one measurement is sometimes the best – this should be taken about 6″-7″ up from the butt rather than the two separate measurements at heel and comb. However, manufacturers quote the previous method and therefore one must use it. To arrive at this measurement have the shooter look directly at you and measure from the underside of his cheek bone to the centre of the pupil of his eye. These measurements run all the way from 1 3/8″ to 1 3/4″. Most individuals measure 1 1/2″ to 1 5/8″. The figures quoted by gun manufacturers for a skeet gun are usually 1 1/2″ at comb and 2 1/2″ at heel *(refer to photo 10)*; this gives a drop of 1 3/4″ at the eye position 6″-7″ up from the butt. Manufacturers' figures for the trap stock, on the other hand, are 1 3/8″ x 1 7/8″; this gives a drop of 1 1/2″ at the eye position 6″-7″ up from the butt.

Ensure the gun is not loaded, then have the shooter aim the gun directly

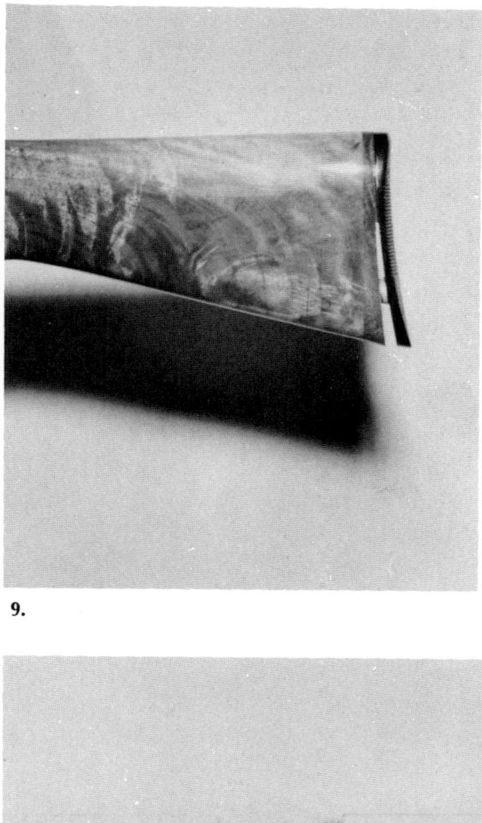

9.

9. How to lengthen stock by adding washers, which also changes pitch. In this photo, pitch has been decreased by adding washers to lower screw between stock and butt plate

10. Gun on measuring board to measure drop at heel and comb and cheek position. Can also be used to measure pitch

10.

at your right eye. His head should be down so that his cheek is placed firmly against the top or, as it's commonly called, **comb** of the gun stock. As there is no rear sight on a shot-gun the height of the comb positions the eye correctly *(refer to photos 11, 12 & 13)*. When the gun is placed on the shoulder and the cheek is hard on the comb the eye looks down or just slightly above the rib. If the comb is too low, the shooter is forced to raise his head and his cheek is not firm against the stock. A low comb on the stock of the gun causes undershooting. Again, if the stock is too high the shooter's eye will be well above the rib. In this latter case the gun will shoot high, or if he attempts to get down on the stock he will have to cant his head over on one side. If he is a binocular type shooter (*i.e.* using both eyes) he'll be in great trouble. With the head canted over and off proper alignment with the horizon, his depth perception, so important to good shooting, will be affected. A one-eyed shooter, on the other hand, will not have this problem and if the comb of his stock is too high, he will be able to compensate by holding slightly under the flight of the target. A shooter using two eyes cannot compensate by this method; he must have his eyes level and the only thing that he can do is to lower the comb of the stock.

Another way to find out whether the stock is the correct height is to have the shooter close his eyes and place the butt of the stock on his shoulder while standing in the correct shooting position. His cheek should nestle snugly against the comb. If he is a right-handed shooter, have him open his right eye (a left-handed shooter should open his left eye). He should be looking directly down the rib or just a trifle above it. If he sees too much of the rib, the stock is too high and if he sees the back of the receiver, the stock is too low. Shooting with too low a stock is worse than shooting with one which is a shade too high. By and large, most shooters use stocks which are too low. Over a period of time they may become good shots but if their stocks fitted them correctly they would do much better, much sooner. In competitive shooting, a shooter tends to tighten up and go down on his stock a little harder – with a low stock he will be in deep trouble.

On pump guns and some automatics, stocks can be raised very easily by

11.

12.

11. Stock too low; shooter's eye below back of receiver

12. Stock too high; shooter's eye above rib

13. Stock correct height; shooter's eye level with rib

13.

 refacing the front of the stock and by bending the tang; on O & U's a little more difficulty may be experienced. High-stocked guns can have the stock lowered or the comb shaved off so that the shooter's eye is in the correct position *(refer to photos 14 & 15).*

"Pitch." The fourth basic measurement of a gun stock, is one of the measurements least understood by most skeet shooters. More often than not, little or no attention is paid to it. Average pitch is often quoted as $2^1/_2$" for a skeet gun with 26" barrels on a pump or automatic or on a 28"-barrelled over and under or side by side. Roughly this means that if the gun is stood on the floor and the top of the receiver placed against the wall with the wall and floor at right angles, the distance between the rib on the barrel and the wall taken at the muzzle, is the pitch of the gun. It is commonly called pitch down; negative pitch is when the barrels touch the wall before the receiver. Guns normally have "pitch down." If you were to compare the pitch of a skeet gun with that of a trap gun you will find that the trap gun has less "pitch down." Trap targets are all rising targets and this requires a shooter to place the shot charge above the target to get good breaks. To achieve this, stocks with higher combs and less pitch are required on trap guns.

The height of the comb on a gun stock contributes more to raising the point of impact than pitch, but pitch is also a contributing factor. All things being equal, stocks with less pitch tend to have a higher point of impact, providing they are shot from the same contoured shoulder. To properly fit a shooter, either "more" or "less" pitch is required, depending on whether the shooter is flat-chested or has a barrel chest. It naturally follows that a shooter with a flat chest requires less pitch and *vice versa.* Pitch can be changed by adding washers to the top or bottom screws, between the butt plate and stock *(refer again to photo 9).* Washers added to the bottom screw decrease the pitch. Careful observations taken by a knowledgeable instructor while the individual is shooting skeet will reveal whether consistent undershooting is occurring. If there are no other reasons such as flinching, etc. for this undershooting, a higher comb and less pitch will enable the shooter to correctly place his pattern in the most suitable spot. Pitch alone may be all

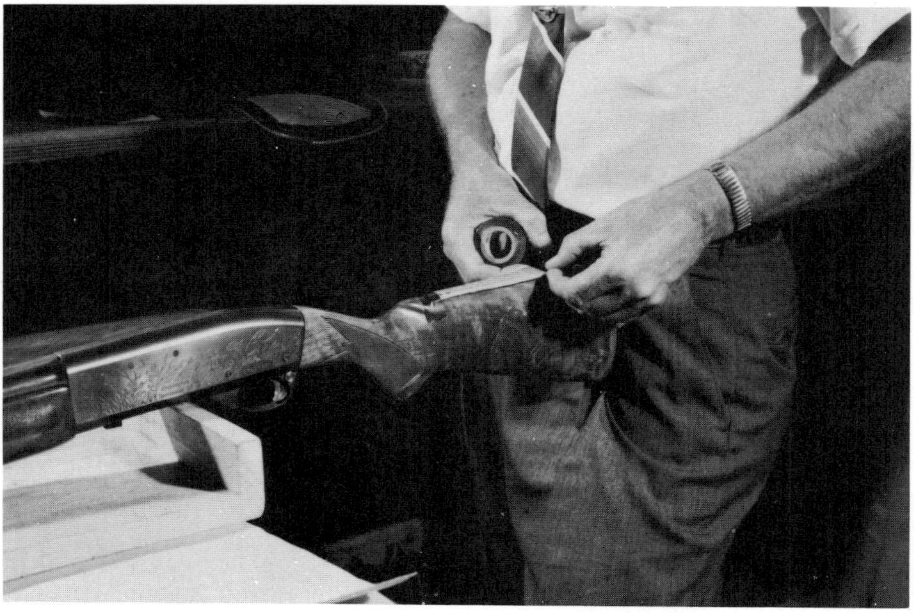

14. Temporary method of heighten ng stock; apply leather and electrical tape one layer at a time

15. Stock here built up layer upon layer; when sure that stock is correct height, have it adjusted by a competent stock maker

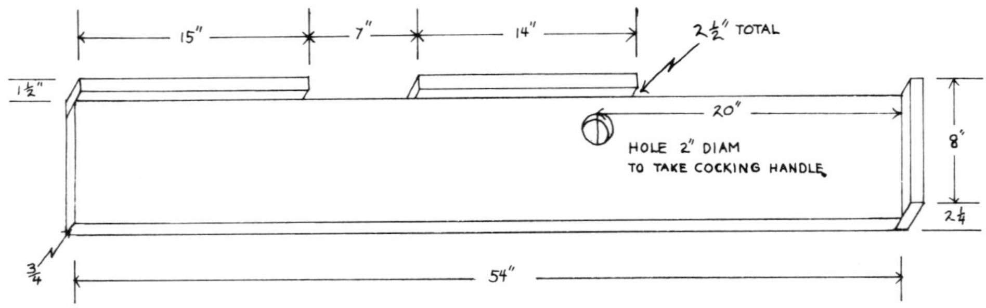

Measuring board for drop at comb and heel and pitch

that is required, or it may be a combination of both. A shooter with several guns should ensure that the pitch is the same on all guns, and also that the stocks are the same height providing that the centres of impact of the patterns are the same when the guns are patterned at twenty-one yards, the ideal distance for skeet. All barrels do not deliver the pattern to the same point of impact. All stocks are not of the same height. The pitch of all gun stocks is not the same either, even though they may belong to the same family of guns and are listed under the same model number.

If you wish to make a measuring board for checking drop at comb and heel and pitch, get a board and make it up as illustrated above.

"Cast." Many shooters bring their guns up to their right shoulder and find it difficult, if they have the correct height of stock, to bring their line of sight to a point through the vertical plane of the rib of the barrel, and, at the same time, keep their eyes level horizontally. To achieve this a shooter should ensure the stock is bent laterally instead of being forced to accept a lower stock. This lateral bend is known as Cast and when the stock is cast so the vertical plane through the stock is to the right of the vertical plane through the rib, the stock is said to have "Cast Off." Conversely, if the vertical plane through the stock is to the left of the vertical plane through the rib of the barrel, the stock is said to have "Cast On." It naturally follows that when this is being considered the gun is held in the shooting position. Generally speaking, a right-handed shooter requires cast off and a left-handed shooter cast on. Shooters with very thin faces may require stocks with zero cast.

You may have seen a shooter, after checking to see that the gun is not loaded, grasp the gun at the muzzle and with the toe of the butt on the floor look directly along the rib. He then slowly lowers the muzzle until the top of the stock comes into view. What he is looking for is the angular bend of the stock and by careful observation he can tell whether the stock is cast "off" or "on."

When speaking of cast there is another term often used and that is "In or Out." Cast "out" is the same as cast "off" and cast "in" the same as cast "on." This may sound a bit confusing. The vast majority of shooters shoot with the gun on the right shoulder so the terminology "off" or "out" is applied directly to them. It is "out", "off", or away from their face. When the right-handed shooter has a gun with cast "in" or "on" it is toward his face. For a left-handed shooter there is no change in terminology but the directions are reversed. The term "off" or "out" for a left-handed shooter is really in toward the face and "in" or "on" is away from the face.

Most shooters require cast off if they shoot from the right shoulder especially if they shoot with two eyes open and if their stock is the right height. Many will say that they do not require cast off; this is true only if they are shooting a lower stock than one which fits them correctly.

Let's take a shooter with a wide chest, shooting with two eyes open and a high stock without cast off. To place the right eye in the correct position in line with the rib, he will have to tilt his head and when he gets his eyes off level with the horizon his depth perception will be impaired. A one-eyed shooter under identical circumstances will have less trouble. Shooting with one eye results in loss of depth perception, so important in skeet shooting. If you suffer from impairment of vision, or if you use the wrong eye as master for the shoulder being used, then by all means shoot with one eye only. If the right eye is the master and it is natural for you to shoot from the right shoulder, then shoot with both eyes open. A thick stock also complicates the problem and has the same effect as lack of cast off. A high stock and cast off usually go hand in hand.

There are other complications associated with correct positioning of the eye and the requirement for cast off. Facial features are important: a fleshy-faced, large-jowled individual will require more cast off. Bone structure is another factor – a giraffe-type neck and sloping shoulders help to complicate the problem.

One of the most important aspects of shooting is the placing of the gun on the shoulder. Some shooters place it very low on the chest and then require a Monte Carlo type of stock; others place it out on the arm and wonder why the black and blue marks appear after a few rounds at skeet. All this leads to only one thing: Find the right spot, and if in doubt seek out a qualified instructor and ask him.

If the right elbow is held high, the right shoulder will be relatively high; by holding the elbow down by the side in a more relaxed position the right shoulder will only be slightly higher than the left. A high right elbow tends to straighten out the shoulder and this presents a flat surface to place the butt of the gun against. A good compromise is to hold the right elbow moderately high at about a 60 to 70° angle from the side of the body. This way you can position the butt on a relatively flat surface *(refer to photo 16)*. Remember to bring the stock up to the face, not the face down to the stock.

I could stop here and say that the gun now fits, but this is not always so. Recoil, ever-present in shot-gun shooting even with the modern automatics, muzzle reducing devices and hydrocoil stocks all may cause trouble. Although trigger-guards and pistol-grips have been designed correctly, a few shooters may find they get bruised fingers. Changing the shape of the guard and the pistol-grip may help to prevent this. Modern guns don't fall into this category, only the older ones do.

If shape and contour of a stock lead to a bruised cheek, reshaping of the stock may also be necessary. A bruised shoulder may be caused by too much drop in the stock, a sharp pointed toe, a stock butt that is too narrow – or even by a shooter shooting with only a cotton blouse or shirt on. Some form of padding on the shoulder is a must. For women quite often a brassière strap with a small metal tab may be a harmful factor. Cast in for a right-handed shooter should also be considered. A remedy for all of these can easily be found.

Gun Balance. A word on balance of a skeet gun. Some guns are muzzle-heavy, others seem to have the weight concentrated in the area of the left hand. In others the weight is between the hands and some are butt-heavy with the weight towards the rear of the gun. For skeet shooting with an O & U, the balance should be through the fulcrum or hinge-pin or just ahead of it. The balance should be spread out along the barrels and fore-end rather than in one place. For skeet, a gun with forward balance is required as there are a great many shots which require gun traverse from left to right and *vice versa*. A gun with forward balance once started on its swing will assist in the follow-through due to the inertia imparted to it (Newton's law). On the other hand,

16. Note position of right elbow; butt of stock hard against the shoulder

17.
Shooter here ready to call for high house target but gun is canted

a muzzle-light gun is easy to start in motion but it has no automatic follow-through and must be propelled continuously or it tends to slow down and stop. A gun which is too muzzle-heavy with a heavy blob on the end has two motions: first, the follow-through; second, the gravity pull. During a long event, a shooter becomes tired and the muzzle-heavy gun tends to roll under the flight-path of the target. With automatics the balance should be through the hole for the extractor with the breech block in the closed position. These positions are for a 28″ barrel over and under and a 26″ barrel automatic with stock lengths of 14¼″ and 14″ respectively. In the case of the automatic, weight may either be added to the magazine cap or placed inside the magazine tube. The O & U is a little more difficult. A larger or longer fore-end with some of the wood removed on the inside and replaced with a mixture of lead shot and plastic wood or glue should suffice. On the other hand, when the stock is too heavy – this is usually the case with a heavy close-grained piece of wood – a lightening hole may have to be drilled to achieve proper balance.

18.
Normal position of gun prior to calling for high house target from station two

What about weight and balance with the smaller gauges? Some shooters advocate the same weight for all four gauges, 12, 20, 28 and 410. Others feel that the weight should drop off as the gauge gets smaller. Some believe that the 12, 20 and 28 should basically weigh the same and the 410 be lighter – more by design than by choice. This is a very controversial subject and I don't intend to take sides. I'll say only one thing and that is, despite the weights involved, a uniform balance from gun to gun is best.

Other forms of shot-gun sport – carrying the gun afield, duck shooting, trap and international trap – present a different set of parameters and by and large they require shot-guns which have different properties.

Gun Canting. From time to time while shooting the high house target on station two, either singles or doubles, a shooter will unknowingly drift into the habit of canting his gun. This rotates the sighting plane on its axis and occurs during the swing back to the ready position *(refer to photos 17 & 18).* The problem is great if the shooter shoots with both eyes open. Right-handed

19.
Shooting six low, gun inadvertently canted; when swinging back to ready position, lowering left shoulder prevents canting

shooters often turn their head in toward the high house and in doing so they take their head away from the stock. In the latter case they never get their eyes level in time to properly see the target.

A similar type of error also occurs on station six low *(refer to photo 19)*. Excessive canting will result in improper pointing. On six low the cant is toward the low house, thus placing the left eye higher than the right horizontally. It isn't until the gun has swept through half its traverse that the eyes will become level. This will result in many "lost" targets. A shooter is not aware of this change and it may remain with him for a long time, either until he drifts out of it or until some knowledgeable shooter, having encountered the same problem himself, tells him. If you are having trouble, examine this aspect; if you see someone else doing this very thing, be a good sport and tell him.

3.

Gun knowledge

Technical knowledge of guns is essential for good shooting. True, there are some topnotch shooters around who don't have this technical expertise. But most consistently topnotch shooters are gun experts too, and most of them attribute at least part of their success to technical knowledge.

First, learn how to assemble and disassemble your guns. Then, with a little additional know-how, you'll be able to carry out preventive maintenence. You should know how to replace an extractor, extractor extension or plunger and spring. Some types of extractors have to be hand-fitted – you may need the services of a professional gunsmith here. Incidentally, if you do have this type of gun make sure you have an extra set of extractors (ejectors) fitted for emergencies. You should also know how to replace a firing-pin and a spring. Once you've mastered these simple skills you'll have the advantage of being able to get your favourite gun back into action quickly after an emergency. Nothing is worse than having to finish a round or a tournament with a strange gun. Good cleaning is essential too; especially so if you're using an automatic. Gas-operated guns accumulate varying amounts of gas and powder residue which may make the action sluggish.

Stock fitting

Poorly fitted stocks can result in the gun severely buffeting the shooter about the face. The cause may be a crooked stock; a stock with too much drop at comb and heel and possibly too much pitch down. This results in the gun driving up and back and hurting the shooter, who may develop a flinch. The solution to this may rest in what is termed cast off.

Cast off

Cast is the angular bending of the stock in towards or away from the shooter. To check the cast of your gun make sure it's unloaded, take it by the muzzle and align the rib with the right eye. Then slowly drop the muzzle away until the top or comb of the stock comes into view. You will see at a glance whether the comb of the stock is in direct alignment with the rib. If the stock angles off to one side or the other you will have either cast on or cast off. What is right for you here depends on your personal build. If you have a round or oval face you may need a slight cast off; if you're thin-faced you may find a stock with a zero cast more suitable. In any event, if you're right-handed the rib of the gun should line up with the right eye without having to turn the head over sideways to properly align the eye.

Once you've found what's right for you – preferably with the help of an experienced shooter – it doesn't necessarily follow that the stock will stay exactly in position. Changes in humidity through the seasons affect wood, and you may find the stock becomes loose in the winter. Or if you've been caught in the rain the stock may be affected after drying out. High-grade guns with stocks fitted onto long projections at the rear of the receiver are less affected than guns with inferior support for the stock.

Trigger pull

Heavy trigger pull or excessive trigger play leads to poor shooting. Many shooters labour under this handicap without realizing what havoc it plays with their scores. Heavy pull can also lead to flinching, which I'll discuss separately later on.

Factories have a habit of setting trigger pull abnormally high – often up to seven or eight pounds' pull is required to release the sear. To overcome this heaviness, some shooters shorten the coil trigger spring by cutting it. This may appear at first glance to be satisfactory but I don't recommend it. It is much better to hone the mating surfaces of the sear and trigger or to bump the trigger spring. These techniques, however, may be a bit beyond the skill of the average shooter; if you're at all doubtful consult a competent gunsmith. But for those who like to tinker around, here are some tips on how to do it. Honing the surfaces will only reduce the tension slightly, but it will smooth out the trigger pull generally. Bumping is a process of reducing the compressibility of the spring by shortening it. The spring is placed in a short section of tube or ferrel, the internal diameter of which will just accommodate the spring. It is then struck a moderate blow using a suitable punch and hammer. To adjust trigger pull properly use a trigger pull gauge – four pounds is the recommended pressure. After setting the trigger pull, shoot the gun to make sure it's satisfactory and that there are no problems when shooting doubles. If you're a multi-gun man, make sure all your guns have triggers set to give the same pull.

Trigger Play

Play is the distance the trigger moves before contacting the sear. There should, in fact, be no play whatsoever; some guns may require a small amount but this should be reduced to the bare minimum. A trigger should be fully engaged and provide a crisp pull, breaking without drag. There's nothing worse than a lot of play before take up, followed by a long trigger pull in addition to drag.

Chokes

The theory of chokes is a subject in itself; I'll restrict myself here to just a brief comment on the relationship of chokes to the optimum pattern for skeet.

A choke on a shot-gun barrel is simply a constriction of the bore at the muzzle. Its job is to confine the shot charge, thus increasing the number of pellets in the pattern at any given distance from the gun. This is necessary for a trap gun or a gun used for hunting at ranges of thirty to forty yards. Skeet targets, however, are taken at approximately twenty-one yards – the distance from any shooting station to the crossing point. Some shooters may take a little longer to get on to their targets and thereby increase the distance to twenty-five yards. But as skeet is a game where it's imperative to get on to the target and fire quickly usually the distance doesn't exceed twenty-one yards – often it's less.

For a 12 gauge skeet gun the choke or muzzle constriction should be approximately .004 inches. This is usually called a skeet choke. Different models may differ in shape but the end product – a hole-proof pattern or uniform density measuring at least thirty inches in diameter – should be the same. Often, though, manufacturers and gun writers will quote this as 30-35 per cent at forty yards. This doesn't help the skeet shooter who's primarily interested in what percentage his gun throws at twenty-one yards. Bearing this in mind, a pattern with a percentage of 65 to 75 at twenty-one yards is considered suitable for skeet. You might think that 10 per cent is a lot – but think of it another way. Let's say 70 per cent (± 5%) and don't forget, this is with number nine shot *(refer to series of photos 20-27).*

Back in the early days of skeet shooting, guns tended to be bored too tight. Sure, it may look nice to smoke the targets when shooting from stations three, four or five, but all it really indicates is that the shooter is working under a handicap. Targets broken into many hundreds of small pieces count just as well, and a gun bored to do just that is what's required. If you have added a muzzle device, make sure you fire test patterns to find out what choke or setting gives you the best pattern. Obviously, this should be done at twenty-one yards. In addition to finding which choke or setting is best you'll also be able to find out whether the pattern is properly centred after the choke device has been installed.

Compensators should only be installed on gun barrels with ventilated ribs. If you do decide to install a choke or compensator on a plain barrel you'll have to bend the barrel or the compensator to centre the pattern. Otherwise, the gun will shoot below the line of sight. At twenty-one yards this can add up to as much as eight inches below the line of sight – in other words, the effective pattern above the line with our thirty inch hole-proof pattern could be as little as seven inches. To get good breaks with a gun such as this the muzzle device will have to be sighted high on the target flight-path.

20. Roll of 4 ft. paper used to check patterns

21. 30″ diameter cardboard circle used in making pattern sheet

22. Inscribing bull's-eye, 30″ circle and quadrants on pattern sheet

23. Pattern sheet properly inscribed

24.
Stand to hold 4′ x 4′ cardboard to which pattern sheet is fastened

25.
Placing cardboard in position

26.
Using car fender as rest to ensure steady aim

27.
Typical skeet pattern — 12 gauge. Black marks indicate each pellet-hole

4.

Guns– speaking personally

Three guns which are quite popular for skeet are the pump gun, the semi-automatic and the over and under. On the pump and automatic a 26″ ventilated rib barrel is recommended, while a 28″ ventilated rib barrel is a prime requisite for the O & U.

What is the best gun for skeet? The answer is simple: The gun with which *you* shoot best. Topnotch shooters don't care what their guns look like as long as they produce results; I don't subscribe to the school which says you must have fine-grained wood in your guns. Here's an example – the other day I was choosing a gun for a friend. I lined up nine identical model 12 gauge skeet guns and looked them over. I finally chose the one with the poorest grain and the lightest type of wood. Why? Because the gun balanced best; high-grade wood is usually denser and heavier. It makes the gun butt-heavy which is not what you want in skeet.

As to the make of gun which I prefer I'll go right out on a limb and say that my favourite is the now-obsolescent Remington Sportsman S-58. Of all the eighty-odd guns I've gone through in twenty years of skeet shooting I've never found a better pointing gun. I would like to add that should a better gun come along I'll be quick to use it. I've just a few reservations, though, about the Sportsman 58; the magazine tube must be properly adjusted to the receiver. It must give parallel planes on the flat of the top of the receiver and the rib of the barrel; if it doesn't, the gun's fine pointing qualities will be lessened. Its sighting too can be affected if the rib and the flat on the top of the receiver are not in line laterally.

Another question I'm often asked is: Why do I shoot so many different types of guns? I use a Remington S-58 12 gauge, a Browning O & U 20-28 gauge and the obsolescent Winchester Model 42, 410 gauge. I can assure you I don't do this just to please all the different manufacturers – it's just that I find these makes of guns best for me in their respective gauges.

This doesn't mean I don't continue to experiment with other guns. Right now, I have a second 12 gauge Remington S-58 which I use as a spare; last year it was my number one tournament gun. I'm also experimenting with an S-58 and an 1100 20 gauge and an S-58 and an 1100 20-28 conversion. For the

410 gauge I always stick to the Winchester Model 42 with either 26″ or 28″ ventilated rib barrels.

When I first started to shoot skeet in 1946 at Goose Bay, Labrador, I used a Winchester Model 12 solid rib skeet gun belonging to the Army. Shortly thereafter I bought my first gun, a Model 12 plain barrel with an aluminum compensator. The barrel-end had been forged to take the compensator body and this raised the front sight. The net result was that the gun shot six to eight inches low at twenty-one yards and to get good breaks I had to cover the target or shoot high on the line of flight. After reading several gun articles I decided to bend the barrel to correct the fault. I used V blocks, a vice, centring cones, a dial test indicator, a lathe and an abandoned building for my test firing – the outside temperature was –30°F. and windy! I mounted two V blocks on a twenty-four inch piece of three-by-three oak plank, and with the aid of a third block and the vice I bent the barrel. The centring cones, dial test indicator and lathe were used to ensure the bend was uniform in one direction. I test fired on pattern sheets in the abandoned building. My ears rang for days afterwards! After this cut and try method I was now able to shoot in the conventional way. The line of sight and the centre of the pattern were the same at twenty-one yards. I might add here that you should never have a compensator fitted to a non-ventilated rib barrel without having the pattern centred and you should always check it out yourself. I suggest that if you want a compensator or muzzle choke device fitted, fit it only to a ventilated rib barrel and you will have less trouble.

As time went on I decided that I wanted a rib fitted to this gun. But I had such a long delay and so much trouble with customs brokers that I never did get the rib on this particular gun. This long delay without the gun may have helped my progress in shooting skeet for it gave me a chance to borrow an 11-48 Remington. After only one twenty-five straight on the practice field I went right out and won a tournament with it. I immediately got a 12 and 20 of the same model equipped with Cutts compensators which were all the rage at that time.

While I bought the two Remingtons, I had also a Model 12 on order, complete with ventilated rib. My best subsequent score with the latter gun was 334 straight.

In Canada, when I first started in tournaments there was no requirement for a 28 gauge gun but as time went on all four gauges were soon being shot. I got a 20 gauge Browning O & U with a 5/16″ rib and 28″ ventilated rib barrels. To get the 5/16″ rib I had to buy a full and modified, as the 20 gauge skeet guns only had a 1/4″ rib on them. I shot this gun in the full and modified configuration and was able to average 98 at skeet. I proceeded to bore out the chokes and, by using the tapered hones, I came up with a beautiful reverse tapered choke in the lower barrel. This was a cut and try method, hone and pattern, nothing scientific about it. I finished the bottom barrel which had a beautiful bell mouth taper with .004″ constriction. In honing the top barrel I was within 1/2″ of having the two tapers meet: *i.e.* the taper from the inside and the one that I was making when the patterns revealed that I had the same as that produced by the lower barrel. In fact it was a little more open. The top barrel now had .0035″ constriction and a 1/2″ flat type choke. With this gun I set a world record of 99.5, highest in any one year, and have won the year-end shields for four consecutive years as an amateur.

Earlier I'd got a Browning O & U 12 gauge which I shot very well right from the start. By this time I had been changing stocks and now required trap stock dimensions except for length of pull. With another gun, a 30″ O & U Browning, and in the full and modified configuration, I could average 98 plus. It weighed only 7 pounds 10 ounces. By this time I was well advanced in know-how with boring and honing of chokes so I soon had the 30″ full and modified opened up. A good friend of mine, Ray Ashton of Ottawa, Ontario, made special tapered hones so that I could play around with gun chokes. This 30″-barreled gun gave me a long run of 1111 straight in registered tournaments and shoot-offs associated with these tournaments. It was the type of gun that I would go to a tournament with, shoot the one hundred or two hundred straight, and wait around to see whom I had to shoot off with. One day, however, at my own club I lost a low five to bring the string to a close.

With the help of Ray Ashton, to whom I'm deeply indebted, I was now able to use this 20 gauge Browning O & U in the 28 gauge event. He made me a set of short tubes with accompanying 28 gauge extractors. This arrangement has never given me outstanding records but it has been satisfactory. Balance is so important and it made itself felt here. I had a larger fore-end made for this gun and I removed some wood on the inside and added shot mixed with plastic wood. This, along with a trap-type stock and weight added, brought the total weight up to 7 pounds 4 ounces with the balance one inch ahead of the fulcrum. My 12 gauge Browning weighed 7 pounds 10 ounces and it too had the point of balance one inch ahead of the fulcrum. Both these guns swing beautifully.

Now to the question of the 410. I have used the Remington 11-48, Browning O & U (by itself and on multi-barrel arrangements), Winchester 42's and recently the 1100 Remington conversion. They all have good and bad points. The 410 I use for all my tournament shooting is the Model 42. This 410 too has its faults and in its factory configuration it is too light. On light guns such as this I suggest heavier or denser wood in the fore-end and stock. Add more weight to the magazine or the stock. Place the weight in the magazine at the forward end and next to the stock-bolt in the hole in the stock. While it is normal to use a 26" barrel in an automatic or pump because of the long receiver, I find that 28" barrel Model 42 is the best. "Ed" Sedlecky of Baldwin, Michigan (the father of Kenneth Sedlecky, the Federal industry shooter), came up with a novel idea. Mr. Sedlecky fitted an over barrel right over my Model 42 skeet barrel and then had Simmons put on a ventilated rib. Result: Better balance, more weight forward and above all, bigness of barrel, for it was a piece of tubing the size of a 16 gauge barrel that he used. A shooter has only two things in which to judge lead on a target if he uses the pointing-out method: the target and the size of the end of the barrel and rib configuration. The width of the rib enters into it also, but as he is a pointer and not an aimer the diameter of the barrel is the real crux to this problem of judging lead or how much daylight is required to be able to hit the target. Shooters miss most targets with the 410 by being late or behind. The relative distance between target and muzzle results to a great extent from the perspective the shooter obtains by the size of his barrel and his ability to judge distance. At least this is my contention. Now by having an over barrel (large diameter compared to the original diameter of the 410 barrel) you see the same gap between the target and muzzle relatively speaking, with the 12 gauge and the Sedlecky conversion over barrel. My scores have certainly been better and I shoot it with more confidence than ever before. The gun weighs just a little over seven pounds.

I wouldn't recommend taking a Model 42 with a 28″ barrel and then putting on a 28″ over barrel. I did this with my second gun and found that it has a pendulum-like swing. I have since returned this particular gun and have had the over barrel removed. My first conversion was with a 26″ barrel Model 42 and I had "Ed" Sedlecky put on a 28″ over barrel. This, in my opinion, is the best.

If your gun already has a ventilated rib on it and the over barrel is centred equally about the inner or original barrel, you'll find that when the new rib and front sight have been installed the gun will shoot a trifle low. In fact you may not shoot it as well at first. Merely raise your stock and do not be afraid of seeing lots of rib when you shoot. I'm sure you will be amazed at its performance. This modification is one of the latest and best in the field of improvement in shooting with the little gun.

The combination of guns I am using at the present time are the ones I shoot the best scores with, and that is why I use them. But I have other guns I'm working on. A Remington Model 58 20 gauge – in fact two of them – and a 58-20-28 conversion. I also have a Model 1100 20 gauge and two conversions – one a 28 gauge and the other a 410 gauge. I shoot them from time to time making the necessary changes to try to find something better. There is something about the point of impact, line of sight, pattern, gun stock fit, balance and the way they break the target that I am not satisfied with yet, so I don't use them in tournaments.

I have been asked why I don't use a standard multi-barrel set of O & U's. Well, I tried them – not only one set but three or four. Never once did I find a set that I could shoot, with any degree of confidence. This was when I was shooting the 12 gauge Browning O & U all the time. This was at an early stage of my career and I didn't know too much about balance. Knowledge of balance, or the lack of it, was to come later; too late for the multi-barrel guns.

My advice is to shoot with the gun that you are able to do the most effective shooting with and the one that will give you the highest scores and the gun that you have confidence in. Lack of confidence or lack of making a decision has prevented progress in many fields and this also applies to the field of skeet shooting.

If I had not experimented as much as I have over the years I would not be or would not feel worthy of being considered capable of instructing on skeet. You have to have experienced all and every conceivable situation and should have a vast knowledge of the tools of your trade to be able to recognize a particular situation and be able to offer corrective action. To do this you must experiment and when you stop learning or picking up new ideas you are then at a standstill or are on the long slide down hill. To quote a recent case, I have always used an anti-fog liquid for my glasses and it is quite effective. During the month of May, 1966, I was discussing the subject of fogging of shooting glasses and a doctor friend of mine, Dr. Gaspar from Malone, N.Y., asked me if I had ever tried soap. I didn't show too much interest and on a chance second meeting with him he persisted. We went into the washroom and he gave me a practical demonstration. In the operating room the Doctor is troubled with fogging of his glasses and by applying a soap film to the glasses and after wiping it off, there is no tendency to fog. You may or may never have had this trouble, but in hot humid days with no wind I often have experienced fogging. So you see you are never too old or have never been around too long that you can't learn something new.

Recently I have been asked what I do to keep in shape and how or what types of exercise I take to properly tone my muscles and body. The answer that I give is, "If I feel like exercising, I lie down till the feeling wears off." Now all joking aside, I am fifty years of age, have led an active life, eat regularly and often but not to excess, weigh 180 pounds and stand six feet tall. I work five days a week and on week-ends if I shoot 10-15 rounds of skeet I walk at least two to three miles just around the skeet fields and I walk a few blocks to and from work each day. I am therefore out of doors two days each week and more often three, by shooting a little in the middle of the week. I don't swim as I feel that while swimming is a wonderful exercise for swimmers it is no good for skeet shooting. I don't lift weights over my head because I have no aspirations to become a weight lifter. don't smoke or drink if that has anything to do with being healthy, and really I am not healthy, although the doctors say that I am according to their standards. My fourteen-year-old daughter can run rings around me in a 100-yard dash so I don't run for exercise. But give me a skeet field, a gun and turn me loose and I can stay with the best of them. Now the question is, what type of exercise is required to be in shape for a particular sport? You tell me.

I will say this, I refrain from lifting because I have a disc problem. If I mow the lawn, my shooting is affected for a day or so. If I shovel snow the results are the same, so I use a snow blower or if there is too much snow I hire someone to shovel it. If at my age I am asked to push a car, I reply by saying, "I'll drive, you push."

I think you should exercise but everything should be taken into consideration. Eat less, drink less liquor, keep regular hours, enjoy nature's blessings (only in moderation) but above all be honest and truthful, and live the way you feel is right and just, for your mind must also be at ease. All these things will make a healthy body and mind and you will be around a long time to enjoy shooting skeet.

5.

To load a shot-gun

At a recent tournament I noticed one shooter carried all his shells in the left pocket. When loading he held onto the gun at the pistol-grip with his finger through the trigger-guard and dropped the shell into the shell port. He closed the gun by pressing the carrier release with his left hand. This, of course, can result in an accidental discharge. There is a correct way to load a shot-gun and this is it: Carry the shells – or at least most of them – in the right pocket. Hold the fore-end with the left hand and nestle the stock at or near the pistol-grip on the right hip. With the right hand, take a shell from the right pocket, drop it in the shell port and press the carrier release with the index finger. The same applies to the pump gun, only this time return the right hand to the pistol-grip and hold the grip with all four fingers and thumb. Close the pump action. If you were to load from the left pocket using the left hand and you weren't too careful where the index finger of the right hand was, you might discharge the gun when you slam the pump action closed.

If you use an O & U, carry the shells in the right pocket and grasp the stock at the pistol-grip with the whole hand. Drop the shell into the barrel. Now close the gun – there should be no tendency to have the index or trigger-finger inserted in the trigger-guard.

When instructing new shooters be sure to start them off correctly. A habit once formed is hard to break. Sometimes shooters walk off the station carrying the whole weight of the gun on the trigger and trigger-guard. This does not look good and is simply showing off. We've all seen Western movies where the six-shooter is twirled on the trigger-guard – this may be all right for a cowboy but don't do it with your shot-gun.

6.

Shooting equipment

For summer shooting, I use a light, ventilated type of vest. A suède leather pad with an Insert-A-Pad is best. Two pockets for carrying shells are a good idea – if you carry most of your shells in the right pocket and a few extras in the left it'll help keep the shoulder-pad pulled smooth when you mount your gun *(refer to photo 28)*. For shooting in cooler weather, I recommend a properly designed jacket or sweater; again, a leather pad is essential. Some shooters even go as far as to insert underarm zippers; they shoot with them open to get greater freedom (*refer to photo* 29). For winter shooting you may find a down-filled jacket adequate. I find if I wear a heavy roll-neck sweater and a light pullover parka this gives maximum warmth plus the necessary freedom of body movement *(refer to photos 30, 31 & 32)*. Finally, as shooters sometimes have to shoot in rain, a large-brimmed peak cap is advisable to keep water off your glasses *(refer to photo 33)*.

Remember, glasses should *always* be worn, whether you normally wear them or not. You need them for protection against broken target fragments; this happens mainly on stations one, two, six, seven and eight, and there's always the danger that pellets may be deflected by a target.

One of the best types of shooting glasses is Bausch and Lomb's "Ray-Ban." These come with a small circle of metal in the centre and they measure 3 mm. DBL (Distance Between Lens). For general shooting a "type two" green shade is best; use a type three on very bright days or when shooting over snow or into the sun. Kalichrome (yellow) is best for heavy, overcast days; clear glass for late evening or poor light. Other types are grey 60's coated, G-15's and G-31's. I also suggest using a good tissue and patented cleaner on glasses. I find C-I-L tissue and "Klear-Glass" best. "Klear-Glass" coats the lenses *(refer to photo 34)*; this stops rain droplets from sticking to the lens, and prevents fogging when you go inside from the cold.

If you reckon you're going to be shooting in the rain it'll be worth your while to get a two-piece rain outfit with a hood. If you don't like shooting with the hood up you can pull it tight around your neck to prevent water running down. Rubber bottomed boots with leather uppers are best for rain work, or you may find a pair of light high-top rubbers with zippers are good

28.
Most shells should be carried in right pocket; this keeps shoulder-pad pulled smooth when gun mounted

29.
This jacket gives complete freedom and is especially good for international shooting. Original idea conceived by the author at international shoot in Oslo, 1961

30. Heavy roll-neck sweater and light parka with shoulder patch and two slash pockets provide warmth and unrestricted mobility in cold-weather shooting

31. Lined trousers and insulated boots complete the outfit, together with ski cap and insulated gloves

 32. Dressed and ready for the target

30.

33. Rain outfit with wide-brimmed cap to keep rain off glasses

34. Klear-glass for cleaning glasses and golf-grip to be used when gloves become worn and lose tactility

33.

31.

32.

34.

35.

enough when pulled over street shoes. As for headwear, you'll probably find a good golf cap with a mesh top best for warm weather; for colder days use a ski cap with ear flaps.

Many shooters frown on using gloves; but just as golfers use gloves for better grip so should skeet shooters. Use a good make of golf glove; find the Par-Mate S-65 best for warmer weather, the S-82 best for cooler weather (*refer to photo 35*). After the gloves have become a bit worn you may find you tend to lose "feel." Use golf-club grip, a paste sold in golf pro shops, to restore tactility *(refer again to photo 34)*. In tournament shooting you may find it best to have two pairs of gloves; an older pair for practise shooting and a new pair for competition.

A good leather shoe with heavy *crêpe* rubber sole is best for skeet shooting during the summer months (*refer to photo 36*). Wide shoes are good as they prevent roll. A leather shoe with a heavy rubber tractor-like tread sole is best when only stands are available and shooters have to walk on gravel, clay, or a rocky cinder-type in-field after a rain storm. The clay substance results in stones, broken targets, etc. sticking to the sole of the shoe; by using this particular tread on a shoe and by merely twisting the shoe, rubber contact is made with the cement stand and there's no slippage (*refer to photo 37*). For winter shooting, wear a shearling-lined flight boot, a felt boot with overshoe or an insulated leather boot (*refer to photo 38*).

35. Golf gloves for shooting; summer on left, winter on right. Tradename Par-Mate

36. *Crêpe* rubber-soled shoes for summer shooting

37. This shoe may also be used for summer shooting but is especially good under adverse conditions (wet ground, gravel, clay, etc.)

38. Insulated or non-insulated waterproof boot can be worn spring, fall and winter; has rubber sole

36.

37.

38.

39. Lee Sonic ear valves provide good protection. They have several small inner valves which close when struck by the sound waves of a shot, but do not restrict normal hearing

40. Large ear defenders afford the greatest protection

If you spend considerable time on the skeet ranges, as you get older you may suffer a loss of hearing. A type of ear protector should be considered early in one's career as a shooter. The Sonic ear valve is one type recommended; these fit in the ear and permit you to hear but lessen the noise (*refer to photo 39*). Ear defenders or hearing guards, whichever you like to call them, fit completely over the ears and give the greatest protection (*refer to photo 40).* It must be borne in mind that when using the latter type while shooting skeet you are unable to hear the trap as the target is released. On stations one and two, six and seven, the sense of alertness, that of hearing, is gone and you must resort to eyesight alone at the acute angle of a two high or a six low; this can make this particular shot more difficult. It is recommended that a type of ear valve such as the Sonic be used instead of the ear defender or the type that covers the ear completely. In trap shooting where one is back sixteen yards from the trap, hearing the trap when the target is released is not possible and the larger type of ear protector is often used, and to good advantage.

7.

Good form

Exactly what makes for good form? Well, some shooters lean forward like the leaning tower of Pisa. Others crouch down as if they were starting a sprint. Both these types sometimes do remarkably well. But make no mistake about it: these postures *are* a handicap. Just think how much better these shooters would do with good form.

Perhaps good form can best be defined by saying it consists of doing certain things the right way over and over again. In skeet this means co-ordinating a number of different actions. The feet must be correctly placed — so must the hands. The gun must be correctly placed on the shoulder with the cheek nestling against the comb of the stock, and with the head in proper alignment. Your body must move correctly from the ankles up as the target appears, and it must follow through correctly after the shot has been fired.

All of this – call it good form, if you like – is vitally important in skeet due to the speed with which the shots must be fired. A fraction of a second may make all the difference between a hit and a miss. Any unnecessary movement is bad form as it merely adds to the time required to shoot.

Now, a detailed look at what makes for good form. I'm assuming the shooter is right handed – in fact, ninety-five per cent are. The left hand should be placed part way out on the fore-end (*refer to photo 41*). Some shooters hold the fore-end with an almost straight left arm (*refer to photo 42*); this gives excellent control but invariably results in slow speed of swing. This leads to undershooting at sharp-angle targets or a roll-type follow-through on crossing targets. It also leads to shooting under the target if you're the methodical type who fires after the crossing point. On the other hand, the shooter who takes a short left-hand grip has lots of "gidup and go" but poor control (*refer to photo 43*).

When using an over and under gun, the fore-end is shorter. You should grip it so that your index finger can reach out and curl around the end (*refer to photo 44*). With an automatic or pump gun, place the index finger along the bottom of the fore-end but well short of the front end. You can properly position the left hand by selecting a rib post and placing the thumb adjacent to it.

41. The left hand is placed part way out on the fore-end. On a gun such as the Remington S-58 or 1100, the best position is thumb opposite the first rib post

42. Long left arm; excellent control but slow swing. Tendency is to pull down as swing progresses. This also gives a roll-type follow-through

43. Short left arm; lots of speed but poor control. Thumb on left hand opposite solid section of rib.

Note: new barrels on late Model S-58's and 1100's have no solid section, so be careful when positioning thumb

44. Correct position for left hand on fore-end of over and under

The right hand should grasp the pistol-grip with the thumb curled around the forward section of the grip and not along the top. The index finger should touch the trigger at the first joint with the finger curled slightly back. It is most important that the finger should be in contact with the trigger at all times. Holding the finger off the trigger may precipitate flinching. The remaining three fingers should take up a natural position to complete the grasp on the pistol-grip.

As targets are supposed to pass over the crossing point at centre field at a height of fifteen feet, take your dry point (sighting shot) at this imaginary point and then place the gun-butt on your shoulder. If you are the type of shooter who leaves his gun-butt on his shoulder and does not drop it down as you swing in toward the ready position, this dry point is most important. Some shooters sight at a point thirty to forty feet above the marker at the crossing point. As a result they place the gun-butt too low on their shoulders. Conversely, some sight directly at the marker, resulting in the butt being placed too high. All things being equal, the former procedure will result in overshooting or shooting too high on the target, and the latter will tend to cause undershooting. To assume the ready position after taking the sighting shot, merely swing back along the imaginary line of the flight of the target and stop some distance out from the target opening. This position will vary with your ability, your reflexes, keenness of vision and sometimes the background.

If this procedure is used over and over again, round after round, it becomes a habit. Once formed, it will never be forgotten and will eliminate many of the variables in skeet shooting.

If you shoot with the butt of the gun dropped slightly down from the shoulder, the height of the comb is of the utmost importance. There's no time for adjustment after you've called for the target and brought the gun up to the right cheek. The comb of the gun hits the face, properly aligned with the eye, and the trigger is pulled, all in a fraction of a second. Remember to bring the gun up to the cheek, never the head or cheek to the gun. While this process is going on, the eyes have narrowed down on the target and the muzzle of the gun is moving along the flight-path. By the time the muzzle has passed or is passing the target, the gun is on the shoulder and against the face just below the cheek-bone. With split-second timing and good form, a well-smoked target will result.

To achieve all this, the height of the comb must be exactly right so that when it's positioned up, under and against the cheek-bone, the rib and sight will be properly aligned with the right eye. Much can be said for this style of dropping the butt an inch or so. This enables the shooter to concentrate his vision on the target as it appears – that is, more on the target than on the rib and front sight. This 80-20 vision concentration – eighty per cent on the target and twenty on the gun – is ideal. With the gun up, there is a greater concentration on the rib and front sight on a basis of 60-40. Sometimes the percentage is even higher or reversed, so that a shooter is concentrating his vision more on the rib and front sight than on the target. This only leads to poor scores and frustrating shooting.

Quite often in the earlier stages, before you and your gun start operating as a well-oiled machine, you'll find the stock comes up first in a jerky motion; the muzzle dips momentarily and then starts racing after the target. Constant practice in proper mounting and swinging should result in the two movements starting together. As the target appears both the muzzle and the butt start to move: the muzzle along the flight-path of the target and the butt up to the shoulder. They both arrive at their appropriate places at about the same time and the trigger is then touched off. This is good form.

Topnotch shooters are often asked whether they "swing through" or "point out" a skeet target. Some of them use both methods or a combination. These terms and "spot shooting" or "snap shooting" are often applied by shooters to the three types of lead or forward allowance used in clay target or wing shooting.

To hit a target crossing at right angles at twenty-one yards, you must place a shot charge approximately three feet six inches ahead of the target if you wish to break it. It makes no difference whether you do this by coming from behind and firing when your gun muzzle passes the target or whether you establish a set lead and keep your gun muzzle ahead of the target while the shot is fired. There are pros and cons for both styles. Usually a new shooter will start out by pointing out his targets especially on stations three, four and five, both high and low house. He may also do this on two and six, but sooner or later he will be using the swing-through method on two high and six low. As time goes on and he becomes more proficient, he will apply this method to three high and five low. On one, two or three low he may continue to point out these targets for a long time – possibly throughout his career as a skeet shooter. Seven high, six high and five high may also fall into the pointing-out category. On four high and low house targets, a swing-through method is often used when the shooter becomes more proficient; however, many shooters resort to the pointing-out method on these two targets as well.

Many shooters snap shoot seven low. But as shooting straight at the target is all that's required it can't really be considered a snap shot in the true sense of the word. One high is somewhat similar, only the gun is moved in a slight downward arc. This target is often broken by a shooter shooting straight at it. Both eight high and low house targets are usually broken by a swing-through style. It is impossible to point out these targets due to the closing speed of the target and the short interval of time available.

Snap shooting has no real place in the skeet field. Shooters who try to snap shoot are very inconsistent. Once the spot has been selected and the motion started, little or no correction can be made. Snap shooting is sometimes employed, however, in field shooting when game is flushed at short ranges.

Shooters often ask how far they should lead a skeet target. Mathematically, a three low or five high target can be given the maximum lead of three feet six inches. The flight of the target is at right angles to the shooting station providing the shot is made over or near the crossing point. It is also a fact that you can overlead by a third and still hit the target, if you must err, err on the long side. The shot charge is roughly thirty inches in diameter at twenty-one yards with a skeet gun and its approximately six feet long. If you overlead, the tail end of the shot string will still make contact with the target.

Which is best – one or two eyes? And where should the eyes be focused

56 when calling for a target? I've often been asked these questions along with: Which is my master eye? A shooter who, through no fault of his own, finds that he has to close one eye, is considered by many to be at a disadvantage. The axiom that "two eyes are better than one" has never been more true. You can see better with two eyes and proper depth perception – the ability to tell accurately how far away an object is – is a prime requisite for skeet shooting. This important quality is only achieved if you shoot with both eyes open. This is often termed binocular vision.

It is also necessary to have the master eye on the same side as the shoulder on which the butt of the gun is placed. If you shoot from the right shoulder your master eye should be the right eye. This means you can look with both eyes – the right down the rib, and the left going along for the ride so that if you close your left eye the sights will be aligned on whatever you are pointing at. There is an easy way to determine which is your master eye. With both eyes open point your right index finger at an object and then close your left eye. If the finger is still aligned on the object your master eye is the right one. If by closing your left eye you find your finger is not aligned on the object your left eye is the master eye.

If you are acting as an instructor, have the would-be shooter point the index finger of his right hand directly at your right eye as though it were a revolver. After doing this several times you should be able to ascertain which eye is his master or controlling eye. If it is the right, shooting from the right shoulder should present no problem. If it is the left eye and the shooter has had considerable shooting practice using the right shoulder, either a change in shoulder will be required or the left eye will have to be winked or closed entirely when shooting.

There is a third type of shooter who has no master eye; neither the left nor the right eye takes over. This is called central vision. If you ask such a shooter to line the gun up with both eyes open and point it directly at you, you will find that the rib and sight does not line up with either eye. Instead, it's in between. There is only one solution. If the shooter wants to use the right shoulder or if he has already been using the right shoulder, he should close the left eye and proceed from there.

As you progress in the skeet shooting game, you'll realize it's important to be able to see whether a target is travelling along the fixed line of flight or whether it's inside or out. With fully automatic traps coming into more general use you will find there are slight variations in flight, both in elevation and in the horizontal paths. Properly placed targets on the throwing arm of manually operated traps are the most accurate; but here again, trap boys often become careless and targets may vary – especially in the horizontal path. To be able to detect the flight variations of these wayward targets is an advantage for the shooter who is able to shoot with both eyes open. Quite often shooters will wink the off eye down enough and still have binocular vision; the left eye is then slightly blurred but depth perception is not impaired. It must be understood that many skeet and trap shooters who shoot with only one eye open are top competitors. However, if you're able to use both eyes you have the advantage. New shooters should be encouraged to shoot with both eyes open; during instruction the instructor should be able to analyze whether the learner is able to continue, or whether he will have to resort to shooting with only one eye open. Good vision, often referred to as 20/20 or 20/15 is essential. If glasses are required for poor sight the prescription should be incorporated into proper shooting glasses. People with poor vision may feel they are under a decided handicap and may be reluctant to start shooting skeet. Actually, it isn't a handicap at all – there are hundreds of shooters with only fair vision who shoot very well. As we get older, eyes change and while we may have had excellent vision in our youth there comes a time when we should wear prescription glasses. Quite often people resent this and it's hard to persuade them to change.

Most topnotch shooters agree that it's impossible to see the target if the eyes are focused on the trap opening. The target travels close to ninety feet per second as it leaves the trap, and before the eyes can focus and narrow down it will be some few feet from the opening. I suggest focusing in a general area six to ten feet from the trap opening, using what is called a universal focus. Some shooters pick up the target quickly; others will not be able to start closing in on the target until it is well along the flight-path. Such shooters won't be able to fire at the target until it's reached the crossing point or beyond. Others, with fast reflexes and good eyes, will break a three high target half-way from the target opening to the crossing point. This means they have focused their eyes on the target and moved the gun along the flight-path. On swinging through the target they have pulled the trigger and get a direct hit while the target has travelled only ten to fifteen yards. While they have been doing this, if something should appear in the target path – a target box thrown from a house, or a broken bird threatening to strike the shooter – they are able to withhold the shot. This takes great powers of acuity and fantastic reflexes – both of which are the mark of good form.

Finally, when shooting station eight, high or low target, a shooter may focus his eyes on or very near the target opening; on eight low due to the angle the target can be seen the instant it clears the face of the trap opening. With the eyes at universal focus you can see the gun barrel and front sight at the same time that the target is sharply in focus. As I emphasized earlier, the eyes should be focused about eighty per cent on the target and twenty per cent on the gun.

8.

Bedroom skeet

New shooters often ask me how they can practice mounting the gun, dry pointing, assuming the ready position and learning how to groove the proper swing and follow-through. Once you've learned the fundamentals by careful observation or with the help of a good instructor, dry practice at home will help you progress further. This is often called "bedroom skeet." According to the rules for skeet, a gun can be held in "any position which is comfortable or satisfactory to the shooter." Most shooters place the butt of the gun directly on the shoulder, trapshooter-style. Those shooters who do not, however, should learn how to mount the gun, take the sighting shot and drop the butt as they swing in toward the imaginary high or low house to assume the ready position. To complete the practice they should call *Pull* or *Mark* according to the house from which they expect the imaginary target, then mount the gun during target pick-up, swing with the target and pretend to fire and follow through.

Use one wall as the face of the range; starting at one corner of the room proceed in a semi-circle just as though the stations were located on the living-room rug. To carry this a step farther, the correct shoes, jacket, gloves (if used) and glasses should be included. It goes without saying that correct foot position should be practised during this stage of training. Care should be taken to develop the proper swing from the ankles up – not just with the arms, as many shooters do.

9.

Shooting station by station

By now you'll be wondering just when we are going to discuss foot and gun position in relation to the crossing point on a skeet field. As we've covered nearly everything else, it's time to get down to actual discussion of each station in detail.

As in golf, foot position in skeet is crucial. Once you have established the correct position, try to use that position on each station. This will give you a firm basis on which to operate.

Many skeet shooters walk on the station, pay little or no attention to the marker which indicates the target crossing point, shoulder the gun and call for the high house target. Without bothering to change the position of the feet, they then either call immediately for the low house target, having previously loaded two shells, or they pause long enough to load a second shell and then call. This casual approach, once formed, is difficult to break, as resulting scores show only too clearly. Being careless, they rarely ever remove the debris from the shooting station but merely stand on it *(refer to photos 44A, 44B, 44C).*

Station one

The target crossing point should be used as a basic reference. This is the point on a skeet field which targets are supposed to cross approximately fifteen feet above ground. At station one, the feet can be in approximately the same relative position when shooting both the high and low house target. Stand with the feet more or less in line with the crossing point, approximately eight inches apart, left foot forward *(refer to photo 45).* The weight should be a little more on the left or forward foot on the basis of sixty per cent on the forward foot and forty per cent on the rear.

Having adopted the correct foot position, the shooter must now place his hands in the correct position on the gun. The gloved left hand is positioned midway along and under the fore-end with an automatic or pump gun, and at or near the end of the fore-end with an over and under. I find that all four

44A. Careless shooter standing on debris

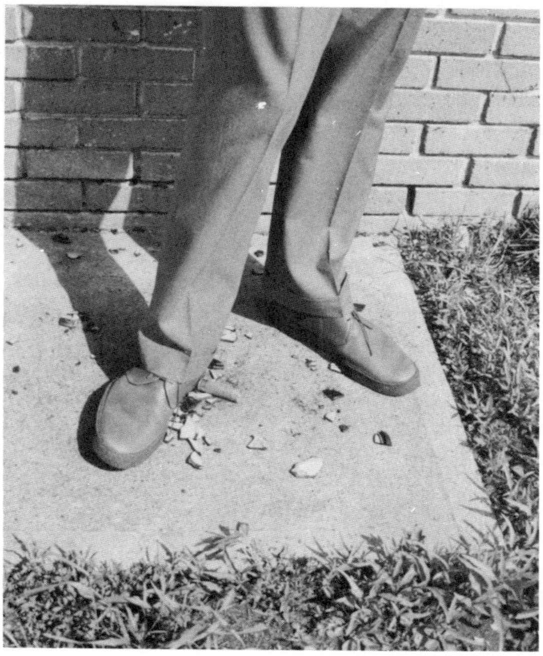

44B. Careful shooter removing debris

44C. Much better shooting platform; no debris

45. Correct foot position for station one high house singles

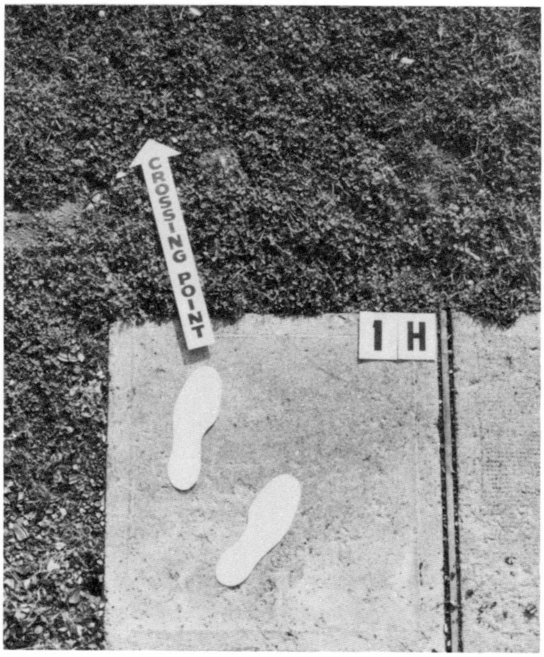

46. Proper position of hands on gun. Note index finger of left hand

47. Proper position of hands on gun. Note how right hand grasps pistol-grip

fingers cannot grasp the fore-end without putting undue strain on the wrist, forearm and elbow. Placing the index finger forward along the bottom of the fore-end is much more satisfactory *(refer to photos 46, 47)*. The right hand should grasp the stock at the pistol-grip with the thumb firmly around – NOT along – the top of the stock. The index (trigger) finger should be thrust through the trigger-guard and should rest on the trigger, making a good grasp essential. Some shooters keep the trigger-finger free of the trigger in case it should go off and only make contact with the trigger when they're about to pull it. This leads to flinching or may prevent pulling the trigger altogether. It often develops after a shooter has had an accidental discharge. To avoid the embarrassment of another discharge, the shooter refrains from getting anywhere near the trigger until he is in the precise act of firing. This type of shooter doesn't know exactly where the trigger is, but moves his index finger in varying amounts each time until the gun is discharged.

Using the crossing point marker as a guide, place the gun on the shoulder and sight over the crossing point to a height approximately fifteen feet above ground level *(refer to photos 48, 49)*. This places the gun correctly on the shoulder and is the prelude to the ready position. Now raise the muzzle of

48. Gun correctly placed on right shoulder; cheek hard on comb of stock

63

49. The dry point, station one. Note the stance and gun position

50.

51.

the gun and you have assumed the ready position *(refer to photo 50)*. Don't raise the muzzle too high or hold it too low; both of these positions will result in lost targets *(refer to photos 51, 52)*. At the call of *Pull*, the target will appear above the gun barrel. Fasten your gaze on it and allow the gun to proceed in a downward arc. Most shooters like to see a little daylight between the target and the rib or sight on the gun. Fast shooters merely cover the target or shoot directly at it.

Having despatched one high, a shooter may, if he so desires, shift his feet slightly around to the left for the second, or low house, target. Again the weight distribution is on the left or forward foot on a 60-40 basis and the feet are separated approximately six to eight inches (*refer to photo 53*).

Again, using the crossing point marker as a guide, place the gun on the shoulder and sight as you did for one high, *i.e.* at the imaginary point fifteen feet above the ground at the crossing point marker *(refer to photo 54)*. Now drop the muzzle down to the target opening on the low house, or just slightly

52.

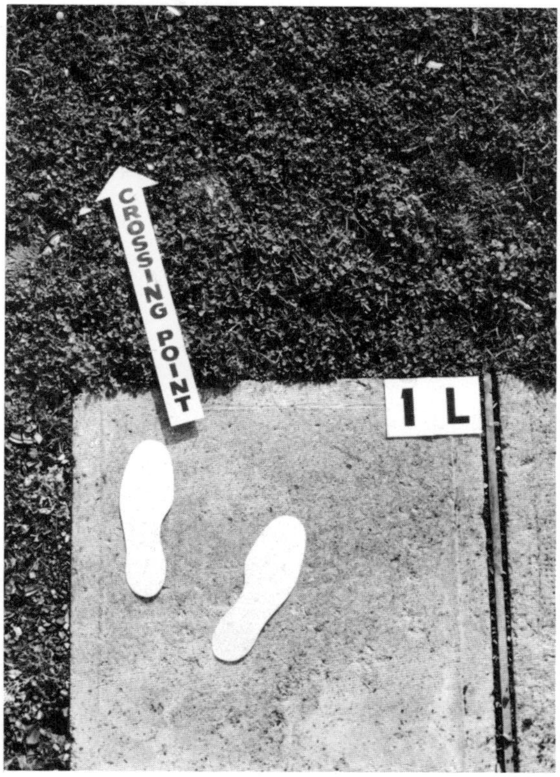

53.

50. The ready position; shooter has dropped gun-butt down from shoulder. Stock is raised to face when butt is placed on shoulder as target appears

51. Incorrect ready position; gun muzzle too high

52. Incorrect ready position; gun muzzle too low

53. Correct foot position for station one low house singles

54. The dry point, station one. Note stance and foot position

54.

 to the left of the target opening *(refer to photo 55)*. Some shooters use the standard word *Pull* for all targets, but the correct way is to call for the targets using the word *Pull* for the high house target and *Mark* for the low. On the low house target a shooter may use either the pointing-out or the swinging-past method to establish the correct lead. A new shooter will possibly like to point out or select a point ahead of the target and keep the muzzle of the gun moving at the same rate as the target. Note that I used the word muzzle. You should point a shot-gun and aim a rifle; in doing this you usually use the muzzle and not the front sight for pointing. This lead of approximately one foot will have to be maintained until the trigger is pulled. The correct follow-through after the shot has been fired will be automatic if you have mastered good form as discussed in a previous chapter.

I'm often asked where the target should be broken when shooting one low. In order to keep the "timing" the same and to take advantage of the maximum pattern density and size, shooters should attempt to break the target over the crossing point. When you watch topnotch shooters in action you will notice that many of them let the target ride in and take it much closer to the gun than recommended. Through years of shooting various habits are formed; while you have been told what I consider the best position, you may find that as time goes on you will break the target well after the crossing point, more or less at the position for doubles from the same station.

You may notice at a tournament shoot that some shooters load only one shell for all singles while others load two shells for all stations except station eight. This may be true whether they are shooting singles or doubles. I strongly recommend loading only one shell during the shooting of singles. This gives the shooter time to adjust or change his stance and to settle down for the second shot. Should he load two shells, the second shot is often hurried without proper care being taken. This may develop into a careless habit. Once the shooter has developed to a point where he is ready for competition shooting, he may find that this habit once formed is not to his advantage. In the interests of safety alone, I think shooters should load only one shell during the shooting of singles.

55. Ready position, station one low house

56. Correct foot position, station two high house

57. Station two high house. Note the imaginary line across the toes towards the crossing point marker

You will note that we started off on a normal round without observing a target or targets from each house for height and alignment. This should be done and I'll explain later the reason for it.

Station two

Again, using the marker at the crossing point, stand with the feet about six to eight inches apart. The tips of the toes of both feet should be on an imaginary line drawn from the centre of the station out to the crossing point. The right or rear foot should be parallel to the rear face of the station. That is, if the station is geometrically correct, at right angles to the imaginary line. The left foot should be placed at approximately a forty-five degree angle to this line. Some shooters may like to take a wider stance and turn the left foot farther out toward the crossing point *(refer to photos 56, 57)*. As on the previous station, place the hands in the correct position on the gun and with the butt on the shoulder assume the proper sighting position. The gun muzzle should point to a spot approximately fifteen feet above the target crossing point *(refer to photo 58)*. Now swing back toward the high house and assume the ready position with the gun muzzle on, or slightly below, the intended flight-path of the target. Leave the gun muzzle out as far as is possible in order to allow you to see the target opening. The muzzle should

58. The sighting shot is taken fifteen feet above the crossing point marker. Note position of the feet

 never be pointing at the target opening. Point it parallel to the face of the high house or at right angles to the plane of the field or walkway between the houses *(refer to photo 59).* It takes practice to find the approximate position. Those shooters who are able to drop the butt of the gun while assuming the ready position have the added advantage of being able to leave the muzzle out farther along the flight-path. At the same time they are able to turn their head toward the high house. Here again the man shooting with both eyes has an advantage.

With the eyes at universal focus you are now ready to call in a clear firm voice for the target. Immediately it is free of the opening narrow down the focus and look well at it. Swing the gun along the path of the target and when the muzzle passes it pull the trigger. With proper timing this will always result in a well-broken target. Initially you may have to daylight the target or actually point it out. This will change, and by using the swinging-through method you'll get best results.

The same general procedures should be followed for station two low. Use the crossing point as a reference, move the feet slightly to the left of that imaginary line and place them approximately six to eight inches apart with a 60-40 weight distribution. The distance apart and weight distribution will depend on each individual shooter *(refer to photo 60).* Some shooters won't want to stand with the feet in exactly the position shown in the photo for station two low house. On the other hand, some shooters may wish to stand farther around. High wind conditions may dictate a modification; observant shooters will be aware of this and make appropriate changes.

With the feet firmly established in the correct position place the gun on the shoulder as you sight over the marker at the crossing point (*refer to photo 61).* Now swing the whole body and the gun down toward the target opening of the low house. Many shooters swing only the arms and not the body. The swing down toward the house should be like the back swing in golf. Again, it must start at the ankles. The muzzle of the gun should be on, or slightly below, the target flight-path and a short distance out from the opening *(refer to photo 62).* At the command of *Mark* (remember, *Mark* is for the low house),

59. Station two high house. Note position of the gun muzzle

60. Foot position for station two low house

61. The dry point, station two low house. Note foot position

62. Ready position, station two low house. Note gun muzzle position in relation to target opening

narrow the focus on the target at the same time start the swing. Again, as for one low, it is best to break the target at or near the crossing point. The pattern is best and the timing will be the same for all shots. If the pointing-out type of lead is used, lead the target by about one and one-half feet if taken at the crossing point. Many topnotch shooters often ride this target and shoot well after it has passed the centre of the skeet field. If you are using the swing-through method it naturally follows that the trigger is pulled as the muzzle passes the target. With the swing properly developed up to this point the correct follow-through will be automatic.

Station three

Now let's go to station three. The stance for three high is the same as for two high *(refer to photo 63)*. The rear or right foot at right angles, left foot at approximately a forty-five degree angle, tips of the toes on that imaginary line out to the crossing point, weight 60-40 on the left foot. Again as for two high, place the gun on the shoulder as you sight over the marker at the crossing point *(refer to photo 64)*. This positions the gun correctly on the shoulder. Some shooters point to a spot over the crossing point approximately thirty to forty feet above ground level. This places the gun too low on the shoulder. Again shooters have gone through these motions and pointed directly at the stake which is approximately two feet high. This positions the butt of the stock too high on the shoulder. The former could result in overshooting and the latter undershooting, all things being equal. *(Again refer to photos 51 and 52 to refresh your memory.)*

Again, how far back toward the high house should a shooter swing the muzzle of his gun? After positioning the gun correctly on taking the sighting shot, swing in toward the high house to a point approximately two-thirds of the way in from the target crossing marker *(refer to photos 65, 66)*. The whole body should swing, and the swing should develop from the ankles. Just think of it as being like one of the motions of the dance the younger set are so keen on, the Twist. The muzzle should be on the flight-path of the target

63. Foot position for station three high house

64. The dry point, station three high house

65. Ready position, station three high house

66. Ready position, station three high house. Note how far back you should swing. Gun muzzle is on the flight-path

or slightly below. When the target appears the muzzle can be located some distance in front of it. By maintaining this distance as the target moves across the field, the trigger can be touched off to get a direct hit. Necessary visual lead or forward allowance is about two feet. On the other hand, if the swinging-past method is used the shot is touched off when the muzzle passes the target. The speed of the swing establishes the correct lead.

The foot position or stance for low house three is approximately the same as for low house two. This also follows for low four, five and six. In photo 67 for low three you will note that the feet have been shifted slightly to the left – this gives more tension on the wind-up toward the low house and results in a better swing and follow-through. Variations in weight distribution and the distance the feet are apart may be required by individual shooters. Before shifting foot position for the low house target, load the gun for the second shot; this pause is essential for consistently high scores. Place the feet approximately six to eight inches apart with a 60-40 weight distribution in favour of the left or forward foot. Position the gun correctly on the shoulder *(photo 68)* and swing in toward the target opening of the low house. Again, how far? A lot will depend on the temperament of the shooter, his reflexes and perception ability, and on the background.

Up to now I haven't said much about the background and this is really the first target that may be affected by it. The high house is usually up and in the clear. The low house targets on one and two are usually shot after the crossing point and also all in the clear. But three low isn't. Unless the field is ideally situated with a perfectly clear background, low three will be a little harder to see. The muzzle should be positioned approximately three-quarters of the distance from the crossing point to the target opening (*refer to photo 69*). Slow shooters – those with slow reaction times – may wish to stay out a little farther and position the muzzle approximately two-thirds of the way. If the field has a perfect background, shooters with keen eyesight and fast reflexes may place the muzzle at a point approximately ten feet from the target opening and *slightly* below. When you put the muzzle that close, you must have it slightly below so as to see the target with an unobstructed view. If

67.

68.

67. Foot position for station three low house

68. The dry point, station three low house

69. Ready position, station three low house. Note the position of the gun muzz e in relation to the target opening and the crossing point marker

69.

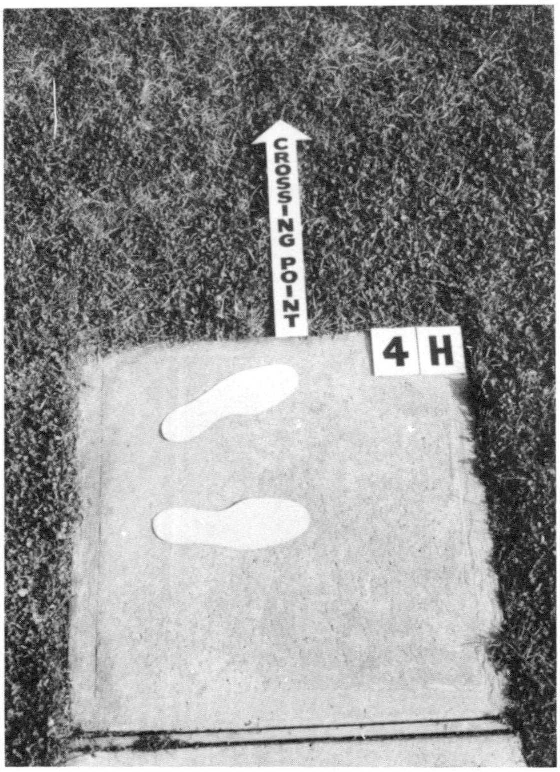

70. Foot position for station four high house

the gun is held farther out it can be held on the flight-path of the target but even then, it's best if it is held slightly below.

Mathematically, the lead for a three low house target is three feet six inches. Generally the lead for this target is said to be from three to four feet. If the shooter uses the pointing-out method, and most shooters do on three low, it is necessary to be able to judge what three to four feet actually looks like at twenty-one yards on a moving target.

Station four

On station four you will note from photo 70 that the foot position is the same for the high house target at this station as that for station three. The distance the feet are apart and the weight distribution are also the same. This repetition, and that of placing the gun on the shoulder and sighting over the crossing point marker, must become automatic (*refer to photo 71*). Practice is the secret of success. Do the same thing over and over again the right way, eliminating all movements that make for a complicated system, and you are left with only the human reaction-time to worry about. On this station the muzzle of the gun can be placed a little closer to the target opening, approximately three-quarters of the way in from the crossing point. Some shooters will find they may like to position the muzzle two-thirds of the way

71. The dry point, station four high house

72. Ready position, station four high house

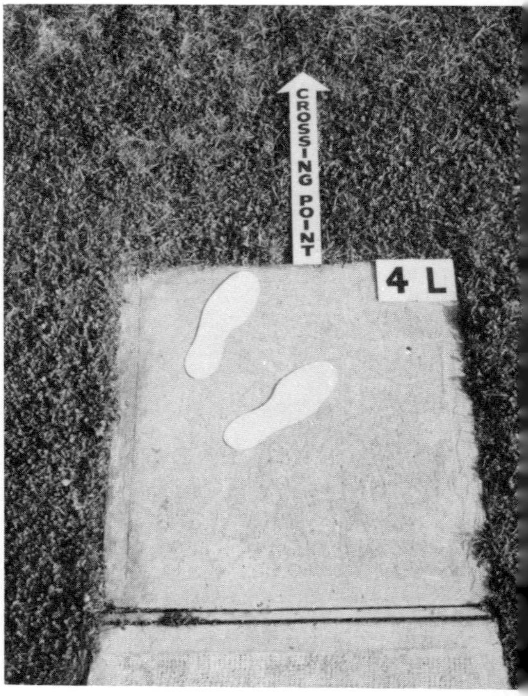

73. Foot position for station four low house

in from the target crossing marker before calling for the target (*refer to photo 72*). This is a variable, and each shooter will find his or her own position. With the muzzle in close to the target opening the target will flash out and the shooter will have to come from behind; this is when the swinging-through method may be used to good advantage. On the other hand, if the gun is started out farther, the muzzle will be ahead of the target and it can be kept there preceding the target across the field until the shot is released. The lead required is approximately three feet.

When shooting four low, foot position, distance apart of the feet and weight distribution are the same as for three low (*refer to photo 73*). Proceed to take the sighting shot (*refer to photo 74*) and place the gun as for three low. Then swing back toward the low house target opening to a point approximately three-quarters of the way in from the crossing point (*refer to photo 75*). Some shooters may wish to go in even farther. When the target appears, establish a lead of approximately three feet and fire. If the swinging-through method is used, fire when the muzzle passes the target.

Station five

Station five high house target requires the same mathematical lead as three low. When shooting at either of these targets the flight-path is at right angles

74. The dry point, station four low house

75. Ready position, station four low house. Note how the body is turned towards the low house. As the body unwinds, the swing is automatic

and if the target is taken at, or near, the crossing point a mathematical lead of three feet six inches is required. Prior to taking the shot the correct foot position, sighting and ready position should of course be followed (*photos 76, 77, 78*).

Low five is almost the same as low four (*photos 79, 80, 81*); the target seems to explode from the opening – you either establish a lead or swing through. You will find that the swinging-through method is best on this shot. Remember, this target starts from a point three and a half feet above the ground and because of the horizon is a little harder to see – so be prepared. One thing must be remembered in shooting station five: do not relax having passed station four successfully. Quite often shooters feel that they are "around the bend." In actual fact they are far from it. In one of the Lordship All Bore Championships a field of one hundred and eighty-seven shooters shot seven hundred and forty-eight rounds of skeet. Five low target accounted for the largest percentage of misses. This was attributed to a letdown in concentration. So be careful!

76. Foot position for station five high house

79. Foot position for station five low house

77. The dry point, station five high house

78. Ready position, station five high house. Note the position of the gun muzzle in relation to the target opening and the crossing point marker

80. The dry point, station five low house

81. Ready position, station five low house

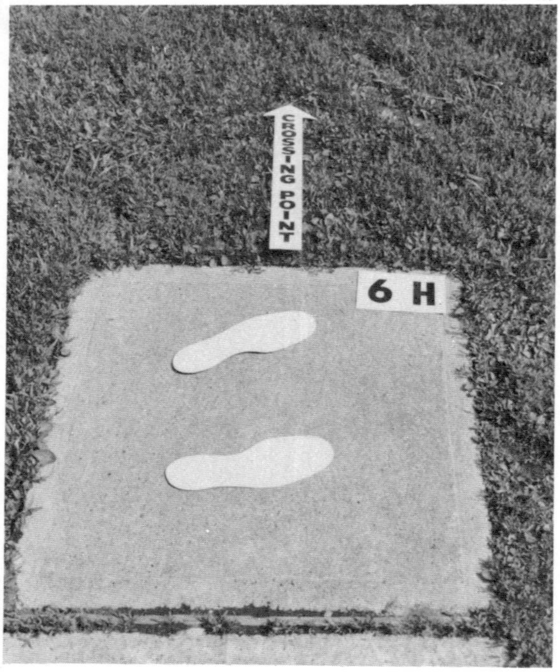

82. Foot position for station six high house

83. The dry point, station six high house

Station six

Station six high house target requires a lead of approximately one and one half feet if taken at the crossing point. If the shooter lets the target ride in, the lead should be adjusted accordingly *(refer to photos 82, 83, 84 for foot position, dry point and ready position)*. Low house six is the reciprocal of high house two. The right-handed shooter will now be able to face the target opening instead of having his back to the opening as on high two. After getting the right foot position *(photo 85)* and taking the sighting shot *(photo 86)*, slide down the imaginary target flight-path. Then position the muzzle part way out from the target opening at approximately right angles to the square of the field, or hold the muzzle slightly to the left of this position either on, or slightly below, the flight-path of the target *(refer to photo 87)*. If the muzzle is held above the flight-path, the target will come out under the gun and the shooter will have a difficult time finding it. By holding slightly below, you will have an unobstructed view and you'll find the target much easier The lead required here is approximately one foot if you are using the pointing-out method. If you are using the swinging-through, fire as the muzzle passes the target. Remember this is a rising target so don't undershoot it.

84. Ready position, station six high house

85. Foot position for station six low house

86. The dry point, station six low house

87. Ready position, station six low house. Note the position of the gun in relation to the face of the low house

84 Station seven

Foot position for both targets taken at station seven is the same (*refer to photos 88, 89*). After taking the sighting shot over the target crossing marker (*photo 90, same for both shots*), drop the muzzle down to a point parallel to the target opening of the high house and just outside the edge of the house (*refer to photo 91*). Most shooters use the pointing-out method and many let this target ride right in. But it is best to shoot it at or near the crossing point with a lead of approximately one foot. Should the shooter allow the target to ride in, the mathematical lead may be reduced. Quite often shooters ride this target all the way across the field. They either become mesmerized and cross-fire or tend to allow the muzzle to fall behind and thus underlead. As a cure, some shooters drop the gun from the shoulder after taking the sighting shot and watch the target approach; they then mount the gun and fire in one motion shooting almost directly at the target.

You'd be surprised the number of times shooters miss seven low. There seems to be no reason for it except carelessness. The shooter is fully prepared, knows exactly where the target is going, yet allows the odd one to get away and is puzzled when it does. This occurs not only in practice but sometimes in tournaments and even in shoot-offs. When it does, the crowd usually lets out a gasp and the shooter feels like crawling into the hole from which the target emerged.

Why? Ask yourself the question. You heard the trap go, you thought you saw the blur of the target as it sped out toward the crossing point, you clutched the trigger and only after recovering from the recoil did you realize that you had missed. One of the main reasons is lack of concentration; seven low looks like an easy target. You become a little careless or you may not stand right. You just about missed seven high as you rode it in all the way and when you fired you could see the target above your gun barrel. Result:

88. Foot position for station seven high house

89. Foot position for station seven low house

90. The dry point here at station seven is the same for both high and low house targets

91. Ready position, station seven high house

You cut the bottom off and just broke it. To break seven low outgoer you should cover the target and fire; it's a rising target, so beware. Many shooters undershoot this target. What often happens is that at the moment of firing, they unconsciously lean forward to catch the recoil, resulting in a clean miss.

If you are shooting in third or fourth position in the squad you have an advantage. You will be able to view a few targets before it is your turn to shoot. Look hard at these targets while awaiting your turn, see how they are behaving and the angle and height they are going. Place your feet correctly for the first shot on seven high and don't move. Load your gun slowly for the second shot and after taking your sighting shot hold the gun without dropping the muzzle and call for the target *(refer to photo 92)*. Fasten or focus your eyes on the target and have a good look at it – let it get out, don't attempt to shoot it too soon. Your most efficient pattern diameter versus density is at approximately twenty-one yards. Now cover the target and pull the trigger. This target is often missed because of shooting too soon. It's nice to see it smoked but it's a dangerous habit.

Station eight

You will note in photo 93 that for the proper stance at station eight the feet are placed a little farther apart for the high house target than for any other position. Place the right or rear foot parallel to the rear face of the station and as close to the rear as possible with the toes about six inches from the right hand edge of the station. The left foot is advanced approximately ten inches and turned toward the high house at an angle of approximately 45 degrees to the right hand edge. The weight is approximately on a 60-40 basis at the start. But during the swing upward more weight may be placed on the forward or left foot on the initial part of the swing and then part of this weight is transferred to the rear foot on the swing back.

Again, as on all stations, position the gun correctly on the shoulder while taking the sighting shot *(refer to photos 94, 95)*. Then drop the muzzle down to a point parallel to the target opening and just to the side *(refer to photo*

92. Ready position, station seven low house

93. Foot position for station eight high house. This is widest stance used on the skeet field

94. Side view, the dry point at station eight high house

95. Back view, the dry point at station eight high house

96. Ready position station eight high house. Note position of gun muzzle

97. Target just before being hit at station eight high

96). Look well at the target as it emerges and swing up through it and as you pass it, fire *(refer to photos 97, 98).* This target is not led. You shoot straight at it by covering it or blocking it off. If you can see the target at the instant of firing you will be late and you may miss. Frequently lost targets result on this station because the shooter allows his gun muzzle to ride behind the target. It is only after he realizes this and starts covering, that good solid hits are made.

Shooters who drop the butt away from the shoulder can start the gun muzzle farther up along the flight-path but remember, don't go too far or you may develop into a spot or snap shooter. Then sooner or later you will find yourself in trouble.

The stance for low eight is somewhat similar to the high house target on this station but the feet are closer together (*refer to photo 99*). Stand to the rear of the station; that is, the edge farthest from the low house and on the side closest to the crossing point marker. The left foot points almost at the low house and the right foot is at an angle of about 45 degrees to the square of the station. The weight is forward on the left foot. If the feet are closer together there may be a slight weight shift in swinging back with the advancing target.

The best spot for taking the sighting shot is in the general area where you intend to break the target; this holds good for both targets at this station

98. A well-hit eight high house target

99. Foot position for station eight low house

100. Side view, the dry point at station eight low house

101. Back view, the dry point at station eight low house

– there is no other reference point available *(refer to photos 100, 101)*. Place the butt of the gun correctly on your shoulder during the taking of the sighting shot and drop the muzzle down to a point just outside or to the left of the target opening. This is the ready position (*refer to photo 102*). Look hard at the opening and as soon as the target appears start the muzzle moving up through it by swinging the upper part of the body and arms back in one motion. This swing develops from the ankles and is the complete co-ordination of the whole body directed and spurred on by the onrushing target. The instant the muzzle flashes past the target the trigger is pulled and the target will disappear in a cloud of dust. This is one station when dust counts. According to the rules, a "dusted target" is one which is usually hit by one pellet and only a spurt of dust is seen. Such a target must be scored as "lost" if there is no visible piece. At station eight when you centre the target in the pattern and shoot it very close, it disintegrates into a cloud of dust and this time it does count.

102. Ready position, station eight low house. Note position of gun muzzle in relation to target opening

10.

Shooting doubles

The stance for one high doubles is the same as when shooting singles. Feet to be more or less in line with the crossing point, separation six to eight inches, weight 60-40 (*refer to photo 103*). Again as on all stations the sighting shot should be taken and the butt placed on the shoulder carefully. I know I've said this before but I've found in years of instruction that this aspect has to be constantly stressed. New shooters pick it up quickly but old shooters have to be reminded time and again. They find it more difficult to get used to the change and only by constant repetition can it be firmly fixed in their minds.

Proper positioning of both feet and the gun are even more important when shooting doubles. The success of the individual shooter will depend to a great extent on his being in the correct position for the second shot. Many new shooters hurry the first shot because they feel the second target will escape before they can shoot at it. This anxiety, while real, is groundless. The second target is usually waiting up there, it even appears larger than the outgoing first – and it's ready for the taking. Having taken the sighting shot for a one high outgoer, shoot it in the same way as in singles. Either daylight it or shoot directly at it. The term "daylight" means simply to put daylight between the target and the muzzle of the gun; this, in effect, is lead or forward allowance. Certain changes may be required if conditions are other than ideal such as during high winds or unfavourable clouds and blue sky. Great care must be exercised in shooting both singles and doubles from this station. One high has accounted for many lost targets.

Keep the face hard on the stock after you have shot at the high house outgoer. Swing from underneath the low house incomer and shoot as the muzzle passes through the target. Both shots are fired in rapid succession and you'll find it is harder to point out the second target. Because of the speed-up in timing, both novice and experienced shooters will use the swinging-through method. A common mistake is to raise the head in an attempt to look for the second target. This is fatal. More often than not you never get your head back down to nestle your cheek against the comb and you overshoot the incomer. Remember to keep the face clamped to the

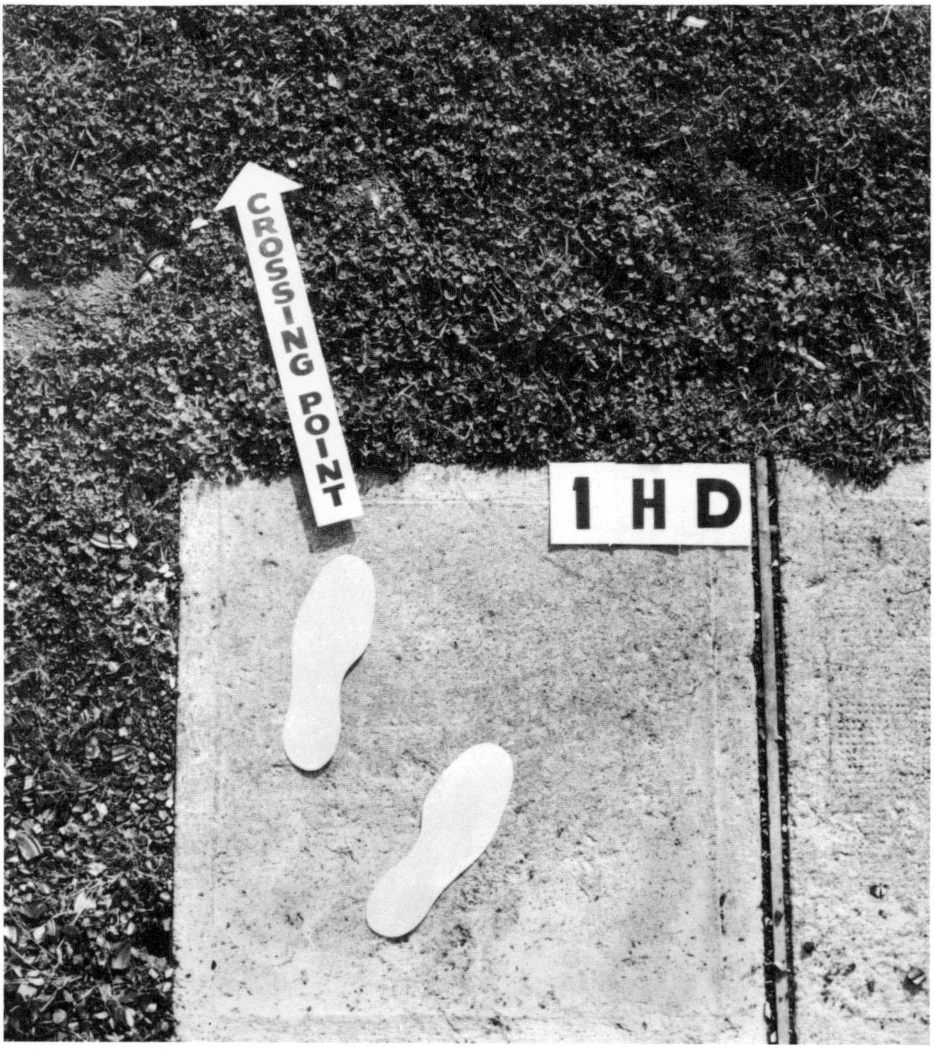

103. Foot position for station one high doubles. Same as for a single one high house target

104. Foot position for station two high doubles. Note compromise between the foot position for a single high and low house target

105. Note foot position in relation to the crossing point for doubles at station two

stock and the eye along the rib. Face and gun should act as a unit. This way you don't get caught "peeking," a term often applied to a shooter when he lifts his head "just to have a look." Due to a speed-up in timing, shooters often shoot their doubles better than their singles.

For doubles at station two you will note in photo 104 that the foot position is slightly different to that for shooting the high house single from this station. To shoot the single target the right foot was placed at right angles and the left foot turned out at a 45 degree angle with the tips of the toes in line with the crossing point. For doubles the right foot is turned slightly and the left is at approximately 45 degrees. If the line is drawn across the tips of the toes that old imaginary line will pass several feet to the left of the crossing point marker (*refer to photo 105*). The reason for this is that the

106. The dry point, station two high doubles

107. Ready position, station two high doubles

shooter must stand in a compromise position to accommodate the second target coming from the low house. Otherwise, the swing to the left will be restricted if a normal stance is taken as for a high house single.

Sight over the crossing point to properly position the gun and swing back toward the high house (*refer to photos 106, 107*). Go after the target when it appears just as you did when shooting singles and keep the face against the stock after the first shot has been fired. Having followed through after the first shot, swing back and as the muzzle passes through the low house target pull the trigger for the second shot. Remember to keep the face against the stock; most low house targets in doubles from station two are missed by shooting over the top.

When shooting doubles from station six, adopt a stance similar to that for

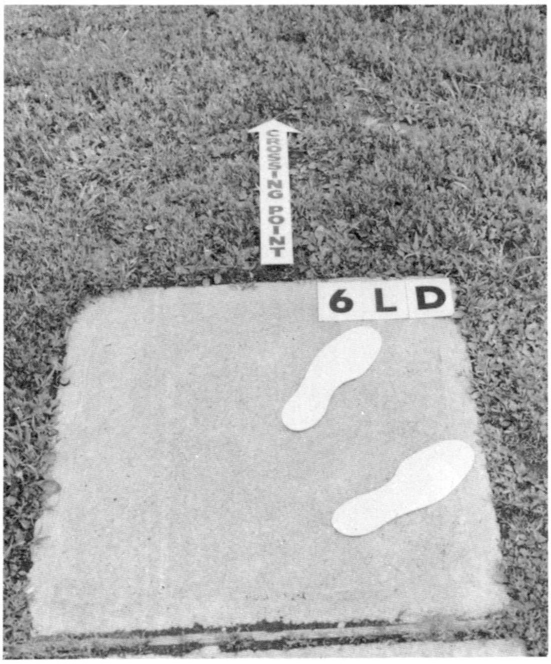

108.
Foot position for station six low doubles. Note again the slight compromise of foot position between a single high and low house target here

six low singles (*refer to photo 108*). Or you may stand in the front right hand corner of the station. Turn the left foot outward to at least 45 degrees to give fore and aft support. If trouble is being experienced on the second shot, I advise standing farther around to the right. But generally I find the normal stance will suffice. The same procedure is followed. Position, sight, swing back and call *Pull* in a firm clear voice. The word *Pull* should be used when calling for the targets at all four stations where doubles are shot. For single targets call *Pull* and *Mark* for high and low house targets respectively. The reasons for this and how to call properly will be covered in a later chapter.

When shooting doubles at stations one and two the outgoer or high house is the first target to be shot at; while at stations six and seven the low house target is the first.

The stance for seven doubles is the same as for singles (*refer to photo 109*). Again, as on all other stations, sight over the crossing point to properly position the gun. When the low house target appears, cover and shoot. Keep the face pressed hard against the comb of the stock, come from behind and underneath the high house incomer for the second shot and shoot after passing the target. The pointing-out method can be used here on the second target but I don't recommend it. The speed with which the shots are fired

109.
Foot position for station seven low doubles. Position is the same as for singles at this station

makes the swinging-through method best. Some shooters flinch just at the instant of firing or fall behind the target when the pointing-out method is used on the second shot. A lost target is the result.

Having gone straight up to this point, the shooter may elect to shoot either the high or low house target from this or any station of his choosing for his 25th or optional shot. Topnotch shooters either take the low or high house target on station seven. Here again do not let down; many optional targets taken at station seven have been missed through lack of concentration. Go through the regular procedure of sighting and swing down to the high house target opening if that's what you choose and call *Pull*. Look well at the target when it appears – either point out or swing through and pull the trigger for your first twenty-five straight. If the low house target has been selected, go through the regular procedure, cover and shoot. Never take this target for granted, be careful but be alert. Don't shoot it too fast but don't delay. Shoot too fast and you may shoot on hearing the trap and get just a fleeting glimpse of the target or it may be slightly off the regular flight-path. Shoot too slowly and you may be too careful and flinch the shot off, momentarily leaning forward in anticipation of the recoil and undershoot. Look well at the target – then cover and shoot.

11.

What makes a good shooter?

You have just completed a twenty-five straight and you ask yourself: How do I learn to go round after round, twenty-five after twenty-five or even put three or four hundred together without a miss. Well, it starts a long way back – some shooters can do it while others can't. It takes time; there are a number of factors involved before a shooter can achieve this goal.

First and foremost is the ability to concentrate before a tournament; this often starts two or three days before leaving for the shoot. You become preoccupied; your family will soon learn to sense that something's different. Your children may say, "Dad is acting kind of queer," or "better watch out, Dad's rather short today," or "he's daydreaming." You can't just turn it off and on like a water tap and this stays with you for a couple of days after you return home from the tournament.

To achieve this concentration you should shoot all your practice rounds just as though you were involved in a tournament. Have proper squads formed; work on each target as though your life depended upon it. Don't fool around making idle talk with your fellow shooters, and before you go out to shoot try to forget all unpleasant things which may have happened. Don't discuss business at the club; don't get involved in controversial subjects; keep your mind alert and think only of breaking targets.

Should you drop a target forget about it; continue to shoot as though you were still straight; be miserable to yourself, if you must, but concentrate. Quite often failure to concentrate may result in a couple more targets being scored as lost due to the shooter's inability to block out the thoughts of the first miss. Don't talk to other shooters during the round but, if you must, discuss only matters concerning the actual shooting – don't bring up unrelated subjects. If someone discharges a gun accidentally, don't jump; just look up quietly as though nothing happened. Some shooters stare intensely at the ground except for the short space of time when they are actually on the shooting station. Others are too busy watching the targets – for example the lead-off man of a squad. He can keep his concentration up by watching the targets of the fifth man on the squad. At station seven he can watch the last high target so that when he shoots at station eight he has a good idea where it's going. Again, at station eight he watches the four high targets because this is the first target of his set of doubles at station one. When he finishes at station two doubles he watches four low house targets, as this will be the first target of his set of doubles at station six, and so on. The lead-off man controls the squad and he can either speed it up or slow it down so that the speed of his squad coincides with the squad on the adjacent fields. The reason for this may not be obvious to the average shooter but it's essential in tournament shooting. There is nothing more disconcerting than to be shooting on station six or seven when the squad on the adjacent field is shooting doubles on stations one and two. Equally disconcerting is to begin shooting doubles on one and two when the squad on the next field is still shooting singles at six and seven – especially if they happen to be a noisy squad. To control the squad properly, the lead-off shooter must have great powers of concentration. The fact that he's preoccupied while shooting helps to prevent his thoughts from wandering elsewhere.

Start your concentration well in advance; have a look at the fields to see how the targets are set. Finish shooting a practice round a half hour or twenty minutes before your flight time. Prepare your shells and don't hurry during the last few minutes. Wash your hands, have a soft drink or chocolate bar and arrive with your squad a few minutes early to have a last look at your first field before the other squad has finished. Study the target flight-path and observe what effect the wind may have. During this hour or so you have been concentrating and you're now well prepared for the shoot. Concentrate during the first round, a little harder during the second, watch the third round and really work hard on it. Be very alert and concentrate on the fourth right up to and including the last shot. By doing this you should have four perfect rounds.

Will to win, desire or sheer determination, whatever you wish to call it, is the second prime requisite for a shooter. At first a new shooter is interested only in breaking targets; after his first twenty-five straight he wants to make it fifty straight. Soon he is shooting well enough and he now looks farther afield. He takes on the club champion – first in fun, then he becomes more serious and he then goes to his first competition. Usually he comes back a sadder but wiser shooter. Taking stock of what happened he attempts to correct his faults and after several competitions he begins to win a few firsts in his class. In shoot-offs associated with tournament shooting you have to show a *will to win.* Work on each target at a time, don't be led into being hurried by a more experienced shooter and by *sheer determination* you may become the champion.

12.

Acuity

The dictionary describes acuity as "keenness of vision." This doesn't just mean good eyes. Sure, superior vision is an asset and corrective glasses are a must for those who need them. But acuity also includes the ability to observe. This starts early in training; if you underlead a target on the same station every time you must know why, and be able to make the necessary correction. It may be the position of the left hand on the fore-end, the position of the feet, the mounting of the gun or the position of the gun muzzle in relation to the target opening when in the ready position. Whatever the cause, a shooter must be keen enough to be able to analyse it and make the necessary correction so that it won't occur again. I assume that you have a correctly stocked gun – if not, this may contribute to constant misses of both end shots and crossing shots. Too much drop is the worst offender; this causes undershooting and even the experienced shooter has a hard time deciding if he shot under the target. Also, the average shooter who rides the target for a considerable distance may flinch the shot. This will also cause the shot charge to pass under the target. Incorrect pitch also has a bearing on the correct placement of the shot charge; too much pitch down leads to constant undershooting. Compensators applied to nonventilated rib barrels will shoot low. Even compensators attached to ventilated rib barrels may not be correctly installed. Proper patterning of the gun after fitting the muzzle device will show whether there has been a deflection of the pattern.

When gusty wind conditions exist you can use the grass, grain or short leafy shrubs that usually exist at skeet fields to good advantage. If the grass is flattened, or if a dark swirly mass shows a dust devil or small cyclonic whirlwind is crossing the field, you can hesitate long enough to let it pass. If necessary you can fumble and drop a shell. By the time you've taken another shell from your pocket and placed it in your gun the conditions will have moderated enough for you to go after the next target. A word of warning here; be very careful if a shell is dropped; leave it there and take a new one from your pocket even in normal shooting. After firing both shells reach down and then pick up the loaded shell. Often you reach down and stand up only to find that your vision or equilibrium has been upset for a second or two.

Watch the flight of the targets carefully; all skeet fields are not located on perfectly flat ground, and there are often slight illusions caused by sloping ground. A target may appear to be climbing while in fact it is going out very flat and the ground is falling away.

It is true that all stations should be three feet square and have two sides parallel to the radius of a circle drawn through the station marker. On most skeet fields this is the case, but you will come across some that are not square with the world. If you line up with the square of the station you may be out ten to twenty degrees. You may say this could never happen; but as recently as 1962 at the World Skeet Championships in Cairo one field had station six out as much as twenty degrees. In the Pan American Games in Brazil in 1963, the skeet field was two and a half yards short of regulation size. This error was caught in time and the field lengthened – but only on the day before the tournament. All practice was on the shorter field. In Cairo the error was not recognized and the whole tournament shot with this bad station.

Large club-houses adjacent to skeet fields can also be a problem. Winds blowing from the club-house to the field may result in unusual air currents crossing the field. If you have a choice of fields, choose those least affected.

On two or three fields at one of the better clubs in North America, the traps have not been levelled for the past six to eight years. When a tournament is held you can bet any money that the high house will curve in, regardless of the wind direction.

Be very observant of the change in light conditions especially if you are shooting after the sun sets. If in competition, be careful what type of glasses are used during this change; an incorrect colour may result in lost targets and the loss of a championship.

Ensure that both hands are properly positioned on the gun and pay special attention to the location of the left hand.

To improve one's score a record should be kept to tell which station is

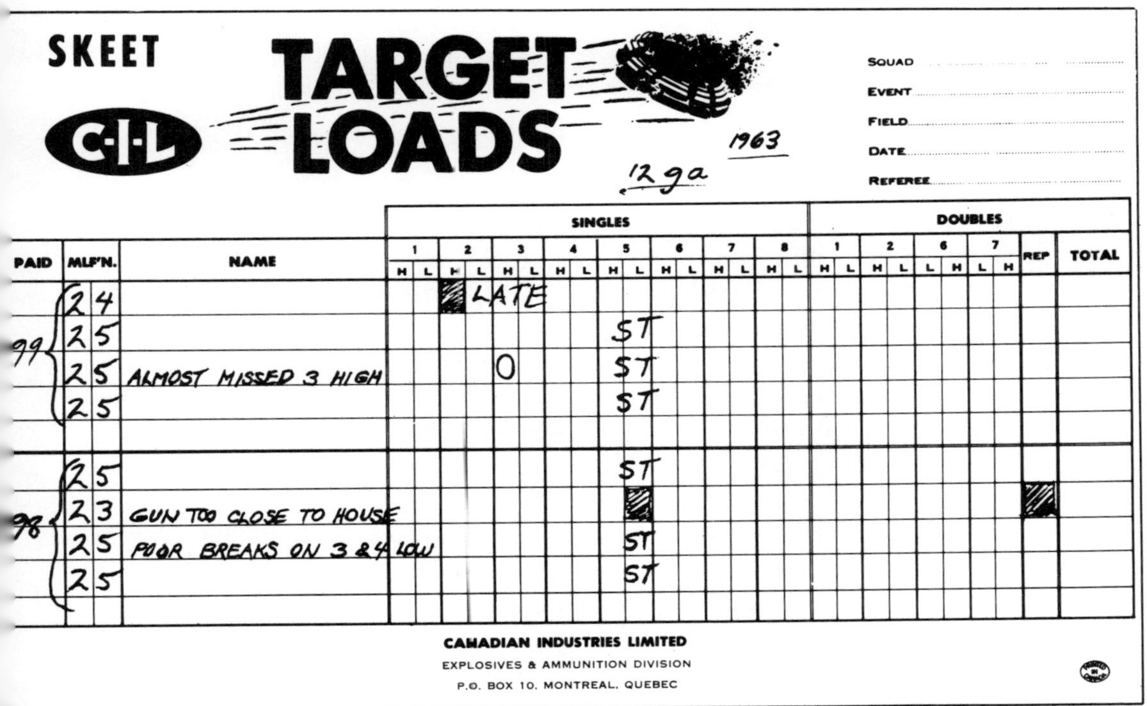

110. Keep a record of your misses; this way you can practice the hard stations

contributing to the greatest number of lost targets. Suitable score pads should be marked (*refer to photo 110*). Block in, in red, the square of the target lost and write in why you think you missed. Four terms often used are *early, late, over* and *under* – these are self-explanatory. Other terms are interference, talking in the squad, slipped on ice, water on glasses, puller slow, etc. Careful analysis on the basis of 100, 500 and 1000 targets and finally a year-end analysis, is the best procedure. If you are a multi-gun shooter and shoot all four guns – 12, 20, 28 and 410 – a separate pad should be kept for each gun.

13.

Stance for doubles at stations 3, 4, 5

 In the 1964 and 1965 Nationals, shoot-offs lasted far too long, especially in the 12 gauge events. When a shooter has to fire 300 to 400 overtime targets it becomes an endurance test. Once you have passed the first two rounds of a shoot-off and you're settled in the groove, so to speak, it could last a long time.

In Virginia Beach at the 1960 Nationals during one particular shoot-off oscillating traps were used to make it more difficult, and to reduce the shoot-off time. Grooved shooters, and those less able to adapt themselves to this new game, found it most difficult. The experiment met with little favour and the idea was dropped. Of course, oscillating traps are also more expensive and this was another reason why they didn't come into general use. A new suggestion to make skeet more difficult and to eliminate so many 100 straights was brought forward. Instead of shooting doubles from stations one and seven it was suggested that doubles from three and five would be better. This, and whether to return to the international rules of low gun-butt and delayed timing, was voted upon by the National Skeet Shooting Association members. The returns showed there will be no radical changes – at least for the present – and skeet will remain as it is.

A procedure already adopted for shoot-offs if the competitors agree is doubles from all stations. This type of shoot-off is quite popular. In regular skeet you already shoot doubles from stations one and two and six and seven and the procedure used for these shots has already been discussed. What is now required is the correct foot position for the shooting of doubles at stations three, four and five.

On station three if you were to stand as for a single high house target you would not have the correct position for the low house target and *vice versa*. It has been found that a compromise is required. Place the feet about halfway between that recommended for the high single and that for the low single *(refer to photos 111, 112, 113)*. It goes without saying that you should take the high house first. You may even start in a little closer to the house, get on the target quickly and break it before the crossing point. There should be a short follow-through when you will be able to pick up the low house

111. Foot position for station three high house single

112. Foot position for station three low house single

113. Foot position for shooting doubles at station three. Take the high house target first

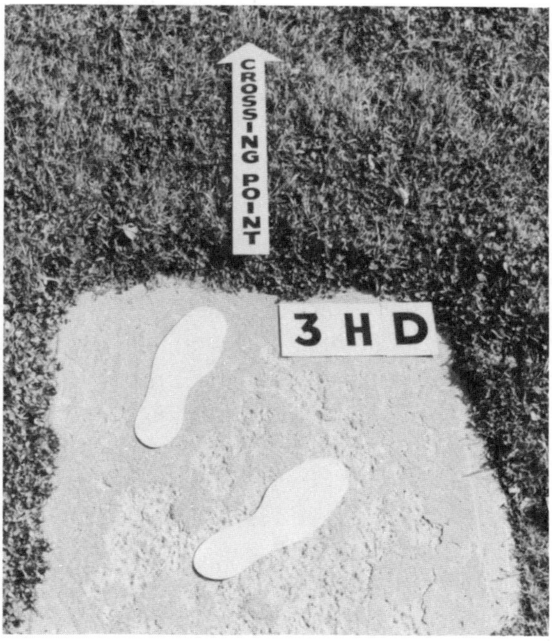

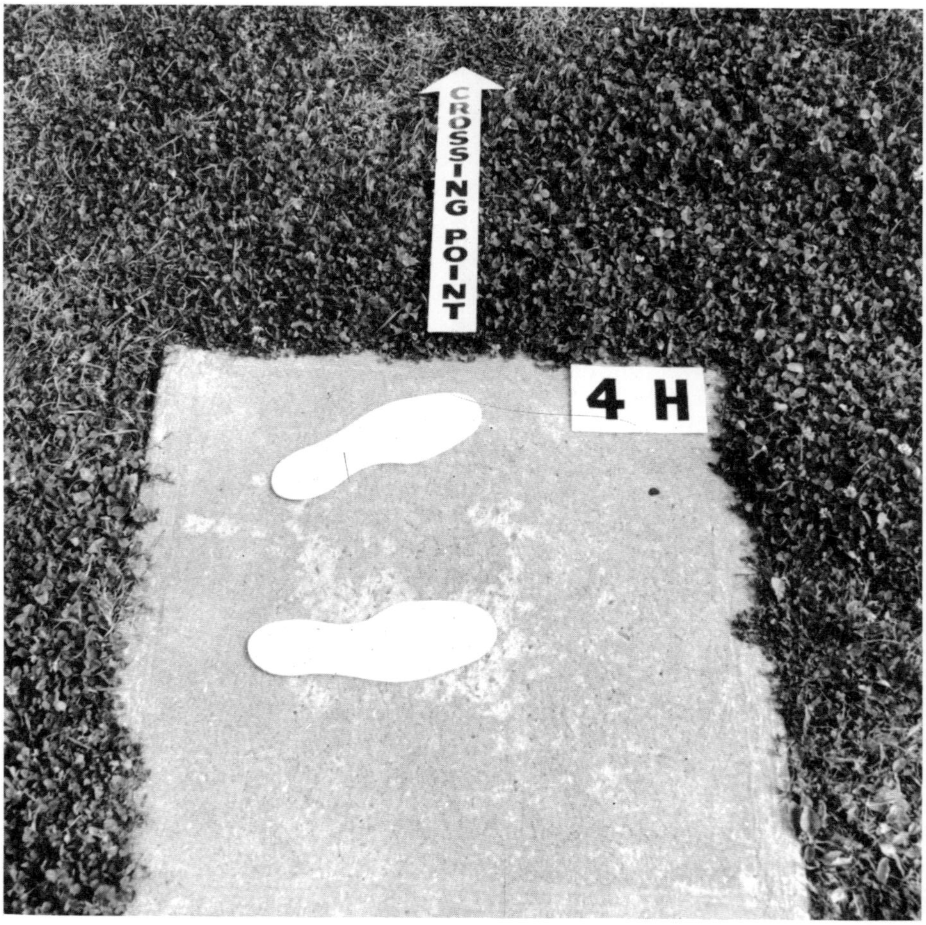

114. Foot position for station four high house single

target, swing with it, get out in front and fire. Don't hurry either shot, the targets are always within range. Many shooters have trouble because they feel they have to hurry; this is not so. Look and see each target individually, not just as a blur or with a fleeting glimpse.

For doubles on station four the same procedure is required. Take up a stance about half-way between the normal or recommended stance for a single high house target and that for the low house (*refer to photos 114, 115, 116*). Establish the proper stance and go through the standard procedure of properly positioning the hands on the gun and taking the dry point. As on three, the swing back on four should be a little closer to the high house than is normally taken for singles. Take the high house target first on station four – if you have some peculiar conditions of light, visibility, high winds or the like, you may wish to reverse the procedure. Many shooters get to station four and if they've never practised shooting doubles on stations three, four and five they don't know which target to take first and they panic. They also

115. Foot position for station four low house single

116. Foot position for shooting doubles at station four. Take the high house target first

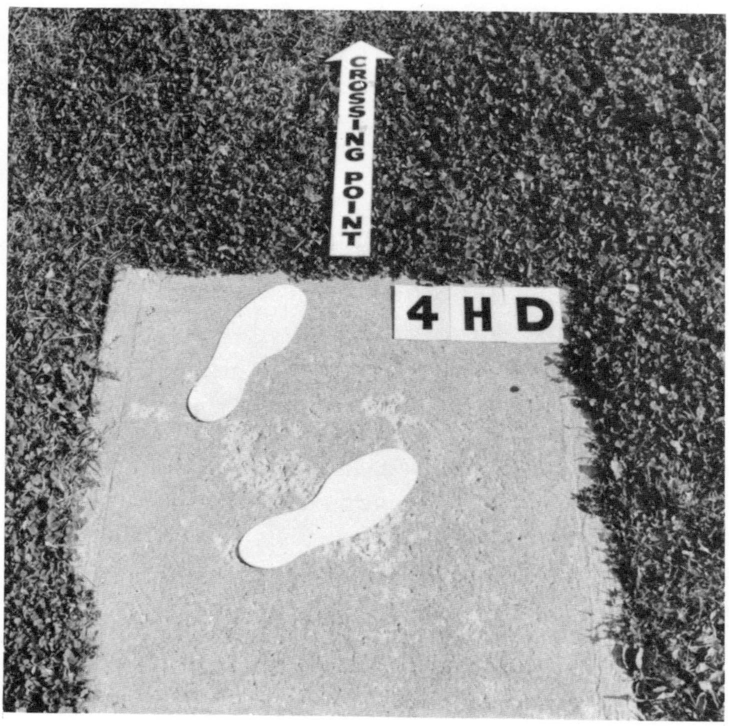

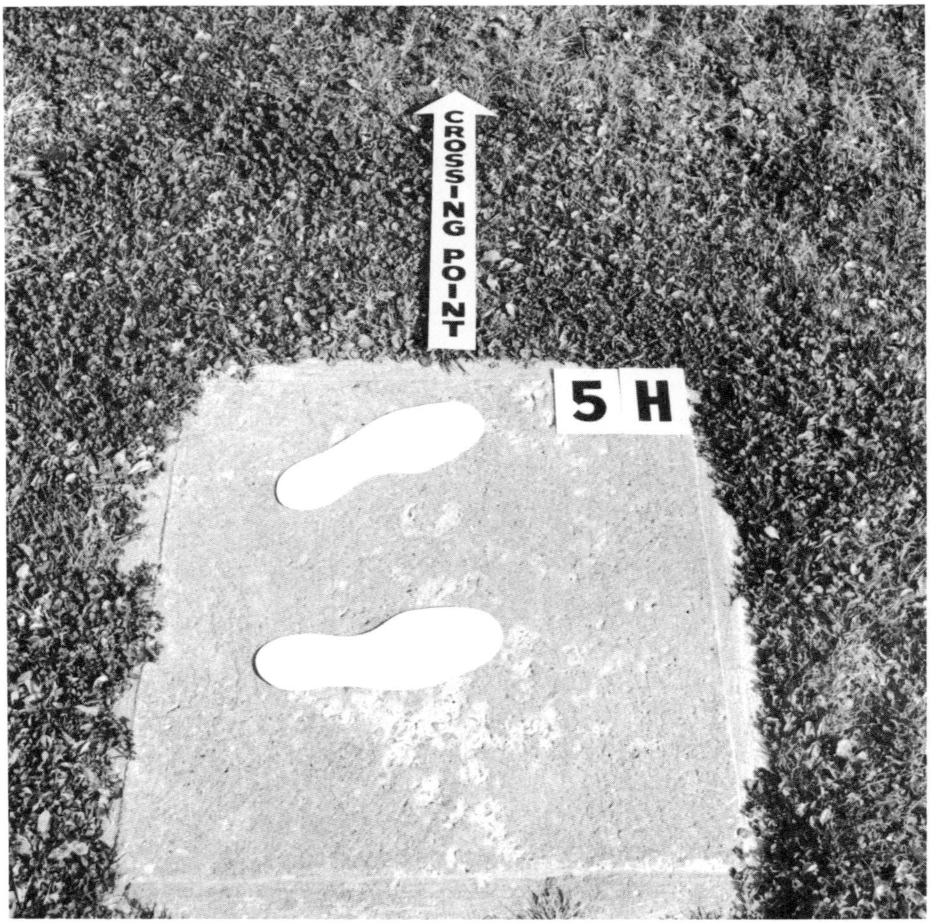

117. Foot position for station five high house single

hurry their shots. You should look well at each target and shoot with good timing. Do not hesitate and do not try to measure each shot or you will fall by the wayside.

Now comes station five. This time take the low house target first and again a compromise foot position is required (*refer to photos 117, 118, 119*). You can favour the high house a bit if you want. Follow the standard procedure of position, hands, dry point and swing back. Swing through the low house target and break it. Get out in front of the high and break it. Look at each target but shoot in good timing. Foot position is the most important aspect of shooting doubles from these three crossing stations. The weight distribution for these three stations is on the same basis of 60 - 40: sixty per cent on the front or left foot and forty per cent on the rear.

If you run into difficulty in getting onto the targets and are habitually late, try shortening up about 1/2″ on the position of the left hand. This gives a slightly faster gun which may be necessary when shooting these doubles.

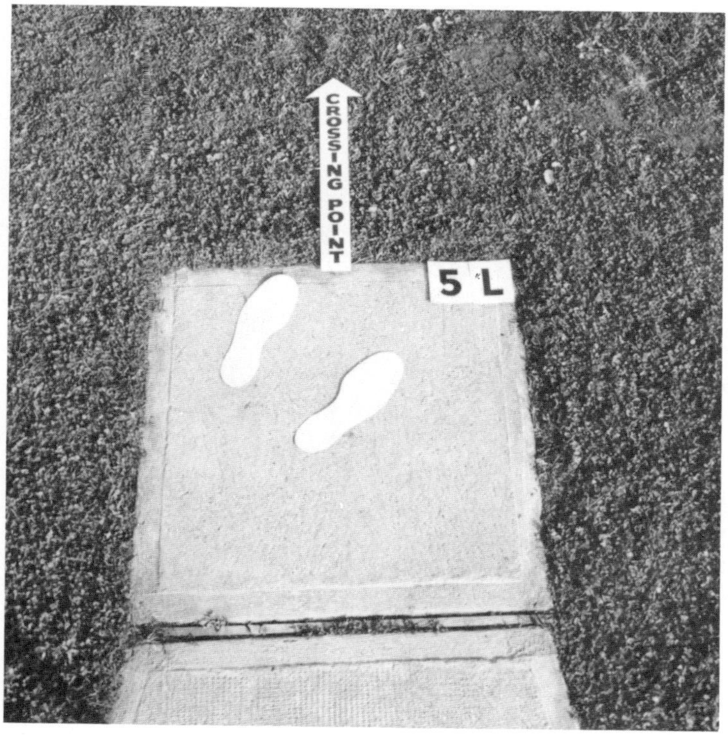

118. Foot position for station five low house single

119. Foot position for shooting doubles at station five. Take the low house target first

14.

How to call for a target

There are half a dozen ways of calling for a skeet target – you'll hear shooters yelling "Hi yi!", "Bird!", "All-right!", "Pull-ll!", "Mark-it!", and, in Newfoundland, simply "Fire!". Actually, although all these are commonly used, there is only one correct way of calling. Use the order "Pull!" for the high house and doubles and the word "Mark!" for the low house. Always call in a loud, clear voice; the puller is watching you and if your call is sharp and clear he'll react immediately. If you call in a soft, quiet voice the target will come out stocking-footed and you'll have a hard time chasing it across the field.

Always give the puller time to get into position after moving from station to station. If you're the lead-off man, give him time to mark the preceding shooter's scores. If a target comes out broken and manual traps are being used, wait until the trap is properly cocked and a fresh target placed on the throwing arm. By calling too quickly you may confuse the puller and upset the squad.

If you are shooting at station eight when adjacent skeet fields are in operation, such as during a tournament, be careful when you call for either the high or low house target. Do not call for your target while the targets are in the air, or when they're in the process of being broken on the field in the direction to which you are looking. It is most disconcerting to suddenly see these fragments and have your target in the air at the same time. Of the two targets, shooting at the low house target presents the greatest distraction. Be aware of what is going on and load your gun; after both shots have been fired on the other field, call and shoot. This way you will never be caught unawares.

If you like to load two shells, do not call immediately for the second target as you will catch the puller off guard. The loading of two shells when shooting singles is not recommended at any time. The loading of two shells is never permitted at station eight.

Should a broken target appear *ask* the referee if you may see another target. The rules permit this, providing you did not shoot at the broken target; if you did shoot, do not ask, otherwise you will present the puller with an unnecessary decision which probably won't be in your favour.

The rule book covers scoring and claims of irregularity where it is associated with "quick" or "slow pulls." By following the suggestions as outlined in "How to Call for a Target", you the shooter, will rarely have occasion to be involved. Bad calls are usually the prime cause of a referee giving a shooter a quick or slow pull; don't be a contributor.

15.

How to start a new shooter

If you have never shot a round of skeet get an experienced shooter to help you. Don't join the first squad and shoot a complete round. It is a mistake to put a new shooter in with a group of good shots and expect much success.

A new shooter should be taken to station seven and, after showing him how to load, let him shoot a few low house targets. Ensure that he places his feet correctly and see that he shoulders the gun properly and that his hands are correctly placed. At this point the instructor may be able to establish whether the new shooter is able to use both eyes open or just one eye. The correct basic fundamentals properly established at this time are very important.

When the shooter has been able to hit three out of four low house targets, switch to the high house while still on station seven. In fact, it is quite possible to shoot the whole box of twenty-five at this station. Explain the purpose of the crossing point marker, ensure that the gun is placed properly and the eye aligned along the rib. Make sure that the stock is the right length so that there is no interference between nose and thumb. Use light loads as this lessens recoil, and ensure that a proper vest is used so he doesn't get a bruised shoulder.

After he has fired several shots, have him walk out of the station and then take up his position on that station again. A 12 gauge gun weighs approximately $7^1/_2$ to $7^3/_4$ pounds and it can become very tiring. Take frequent rests and alternate between the high and low house targets.

On the going away target he will find that all he has to do is cover the target and pull the trigger. On the high incomer, he will have to lead the target a short distance in order to get good breaks. If he stops the gun, a lost target will result.

If the shooter is quick to learn, station one can follow and then stations two and six. When the instructor is satisfied, a complete round may then be shot. For the first day two rounds should suffice followed by similar periods of instruction. It is important that a new shooter gets encouragement by breaking targets and it is most important that he is not permitted to overshoot or *shoot himself out*.

If the shooter is young and small of stature a 20 gauge may be the best gun to start, followed by the switch to the 12 gauge as soon as possible. With the 12 gauge, shooters have a better chance; this *big* gun should be mastered first before using smaller gauges. One prime purpose of the smaller gauges is competitive shooting; this should be the ultimate goal of at least eighty per cent of those interested in skeet.

A new shooter often experiences trouble at station eight. Instead of shooting from the regulation distance of twenty-one yards, move him back three or four yards until he's able to make successive good hits. Then move him forward until he is breaking targets when standing in the normal position.

To create interest if the shooter shows promise, always give him a few shots at the end of the lesson on a new station, the next in line for a subsequent lesson. This whets his appetite and gives him something to look forward to.

16.

What to look for when your scores drop

Shooters in Canada and in the northern United States have a problem: They all shoot in deep snow and cold weather for four or five months of the year. During this time scores may drop off and you may become unnecessarily worried; do not make any radical changes in clothes, equipment, shooting stance, etc. When the warm weather returns, scores will improve. Changes of seasons and heavier clothing affect scores; this can be cured by removing the butt plate when using heavy clothing in winter. With the return of spring you can put the butt plate back on the stock.

Over shooting, that is, shooting too much, especially with the big gun (12 gauge), results in attempting to catch the recoil with the arms. This will cause a jerky type of movement which may result in flinching, and resultingly poor scores.

When using a pump gun, shooters often pump the shell out almost at the same instant the gun goes off; this can result in a jerky swing, and poor scores. Wait until the shot has gone off and take the gun from the shoulder before ejecting the shell; this will break the habit. The 12 gauge causes this more frequently than smaller guns; it can become a bad habit which often affects experienced shooters who shoot a lot.

As you get older your eyes change; this can be dangerous insofar as scores are concerned, especially if you don't realize that a change has taken place. You blame your gun, start trying new ones with higher or lower stocks, and you get into real trouble without properly analysing the problem.

Quite often your gun may be damaged; it may be a compensator or a barrel. Shooters have been known to suddenly start having poor scores with the 410 only to have someone suggest that possibly the barrel was bent. Small bore guns such as these should be given careful attention during transportation in the back of a car and they should be checked periodically by test firing to ensure that the pattern is in the same spot.

Too soft a recoil pad and bad trigger pull may cause flinching (see Chapter 22). These, and a decline in health, can contribute to a falling off in scores. A gun may be sent away for repairs and when it's returned the shooter is unable to shoot it as well. The stock may have been removed and not replaced in exactly the same way. A recoil pad may have been fitted and the pitch changed without the shooter being aware of the difference.

And last, but not least, remember always that it's the desire to win, that makes a champion what he is – take away desire and he's an also-ran.

17.

Etiquette

 The dictionary defines etiquette as the forms, manners and ceremonies established by convention as acceptable or required in society, in a profession or in official life. This also applies to skeet. A shooter should be on his best behaviour at a skeet club, and especially so when actually shooting. This is not only for his own benefit but as an example to younger shooters.

Both in tournaments and during practice rounds be on time when your squad is called. There is nothing worse than having to wait five or ten minutes for a squad member, especially during a tournament. From a psychological point of view the four shooters who are waiting for you to arrive are already at a disadvantage. Some shooters are habitually late; the cure for this is to proceed without them or refrain from squadding with them in future.

If you are the lead-off man in the squad, ask the other shooters if they are ready to observe the targets and ask the referee to pull the low house target or targets and the same for the high. Don't call pull and mark as though you were ready to shoot; this will only cause confusion. "Please" and "Thank you," the magic words, are applicable here too. According to the rules, "At the beginning of each round, when the squad is assembled at station one, the squad shall be entitled to observe two (2) regular targets from each traphouse and shall have the option of observing one regular target after each irregular target, except where said irregular target was fired upon."

When you have shot your pair of targets on the station, walk off without hindering the next shooter and stand so as not to distract him while he is shooting. Do not proceed to the next station until everyone has finished on the station in question.

Conversation should be restricted to the game at hand. During practice and tournament shooting, talk should be kept to a minimum.

Carry your shells in a shell pouch or in the pockets of shooting vests or coats. Nothing is more disconcerting than to have a shooter carry a box of shells around with him and have to pick it up and place it down at each station. When this type of shooter requires an optional shell he has to vacate the shooting stand, retrieve a shell from the box and get back on the stand;

this procedure is most upsetting. If he continues to do it kick over the box accidentally once or twice and he'll soon learn!

Respect the referee and his decisions – do not be a target claimer. If the referee calls **"Lost"** and you thought you saw a piece fall off, you may in the heat of competition call "Dead *bird*" and if he says, "I'm sorry, it's lost," don't argue – break them better the next time.

Know the rules well, study and review them constantly; you may feel free to discuss rules with the referee but always be humble. In competition if you have a gun breakdown, don't try to carry out repairs on the shooting station. Borrow another gun and continue so as not to upset the "rhythm" of the squad. Keep the action of your gun open at all times and load only after you are on the shooting station.

Often when a squad is observing the targets on station one, two or more shooters will follow the target in and dry fire. While this practice may ensure that the safety is off, it can be disconcerting to other members of the squad who are not used to it. Do not aim at a target and follow it across the field unless you are on the shooting stand with gun loaded and have called for the target and intend to shoot at it.

Refrain from giving vent to your feelings by using abusive language when you have lost a target; accept it and proceed with the next shot. Some shooters throw the empty hulls half way across the field, others stand and stare at the disappearing unbroken target or continue to aim at it until it strikes the ground and then turn and look at the rest of the squad hoping someone will break the silence. You have selected a competitive game and the less public display you make the easier it is on all concerned. Hide your feelings as much as possible; it often helps you to win in shoot-offs.

Do not do anything that would upset or distract a shooter, such as accidentally closing the breech of an automatic or unconsciously working a pump shot-gun. If it is late in the evening and the shadows are well out on the skeet field, don't walk around so as to cause the shadows to move in front of the shooter.

18.

Tournament shooting

Here are a few additional tips which may help to round out your education in tournament shooting.

If you go to a tournament which requires signing up on the actual day, a trip to the meteorology office may help you to make up your mind what time of the day would be best for shooting. If rain is expected in the morning, it may pay to shoot late in the afternoon. The best time is from 9 a.m. to about 3 p.m. standard time, but this also depends on how the actual fields are located. Fields with bad background should be studied ahead of time so as to get the best light conditions. Early morning shooting is good for light winds but you may run into having to shoot into the sun. Nothing is worse than shooting at a one high target which disappears into the glare. Similarly, late evening shooting is usually done under calm conditions but again, the sun may present a problem. The diurnal change in wind gustiness usually takes place from noon to 3 p.m., when it's at its peak. While the wind may be a problem the light conditions are at their best at this time. If you become involved in a shoot-off in late evening after the sun has set, be sure to use glasses with clear lenses. Under these conditions a slightly different technique in assuming the ready position is recommended. As dusk approaches, it becomes increasingly difficult to see the targets as they flash out of the trap openings. Starting the muzzle of the gun farther from the target opening but on the target flight-path is the most significant single secret of late evening shoot-offs.

It goes without saying that good health habits are essential and that proper rest is required. Drive to the tournament in daylight hours if possible. If you are with fellow shooters, take turns at driving. If you have to drive long distances alone it will be less tiring to grasp the steering wheel at the bottom with the forearms or wrists resting on the legs, than to grasp the wheel at the top. In this way your arms are not tired out before you reach the shoot.

Some shooters arrive a day or so early and then spend hours at burning up the powder; this late practice is not a good idea.

A few practice rounds of skeet are all that should be required. If you shoot steady for a couple of days and then shoot the small guns, by the time you are ready to shoot the big gun you will be shot out, your nerves will be jumping and you won't be able to concentrate for a hundred straight. Some topnotch shooters say that they never consult the score-board to see what scores have been recorded before they start shooting. As far as they are concerned they are just interested in getting the first hundred straight and then meeting their opposition in the shoot-off later in the day. Don't worry what the other person has just shot; go out and get all that you can.

If you shoot early against a large entry and you have gone straight, go to your motel. Have a quiet dinner, a rest and a fresh shave and arrive back at the field in good time for the shoot-off. Why the shave? If you have a tough beard, by five or six o'clock you will notice a roughness against your gun stock. Quite often a quick shave with an electric razor on only one side is all that is required. Is this being supercilious? Well, it might be – but it pays off in shoot-offs won.

Quite often over a period of time the blueing on a gun becomes worn and the metal may become shiny. The ribs on guns are affected most. Frequently I will re-blue the ribs and the top of the receiver of my tournament guns to present a flat black surface. When I step up to shoot there will be no glint from the sun. This is just another tip from a tournament shooter, but a most important one.

In using the Model 42 once in a while I have a pumping problem and a shell may jam. To help prevent this I often wipe a smear of grease on the breech of the gun, which helps to prevent hang-ups.

19.

Watching the targets

Some shooters watch all the targets that other squad members shoot at. Others in the squad stare at the ground until it is their turn to shoot. Again, some look at only the targets of the shooter in front of them. While not shooting, much of a shooter's time can be spent on watching the flight of the target. The lead-off man should watch the two targets of the last shooter in the squad. Shooting in other than lead-off position, if it's a well behaved squad, is an advantage. Here, you are able to see the flight-path of the targets while standing behind the station before it is your turn to shoot.

After the squad has assembled at number one shooting station, the lead-off man should ask if the squad is ready, then ask for a low house target and observe its height at the crossing point. See that it crosses within the three foot limit and check the general flight-path and where it lands in relation to the fifty-five yard distance marker stake. If automatic traps are in use call for a second target. The same procedure is used for the high house target. Pay particular attention to the height and distance on the first high house target and whether the second target crosses directly over the marker. If manually loaded traps are used, a shooter need only call for one target from each house. The reason you should ask for the high house last is so you will have a look at the more difficult target immediately prior to shooting it.

You may wonder why I suggest that you observe two targets when automatic traps are being used and only one from a manual trap. During squad changes, there is always a delay before the new squad starts shooting and during this time the traps are running and vibrating. If the trap is not perfectly level horizontally the target will either vibrate in toward the rubber rail or out toward the target finger stop. This movement – it can be approximately $^{1}/_{4}$ inch – will give a variation horizontally of the target at the crossing point of approximately two to three feet. The first target from the high house should only be watched for height and distance. The second target, for its position over the crossing point marker, because it will be the most accurate. All subsequent targets during the shooting of the round from the high house will be uniform and compare favourably to the second target that was thrown for observation purposes. This is more applicable to the high house than the low, as the low house trap is at a steeper angle. The vibration has little or no effect in the actual positioning of the target on the carrier arm of this latter trap. As the time element is the major factor, it is only possible to have this occur during a lull in shooting, such as when a squad is changing or when a time elapse occurs during a trap breakdown. During normal shooting, targets are being released too frequently for this to occur. It must also be remembered that if the trap is absolutely level horizontally this problem should not occur. To be on the safe side, always have a look at two targets regardless. On the manual trap it depends whether the trap boy is conscientious and places the target in the exact spot against the limit stops. If he doesn't, anything can happen – and usually does.

Each shooter should check the flight of the targets of the previous shooter and the lead-off man should watch those of the last man on the squad. When standing at station seven while the other shooters are shooting in their turn, watch at least one or two high house targets if you are lead-off man, prior to proceeding to station eight. Again if you are number one in the squad you should watch another couple of targets from the high house, as that will be your most difficult first target of the pair of doubles at station one. One high of this pair of doubles sometimes gets away, so be sure you know the flight-path of this target. To assess the flight-path, stand out in the middle of the field after you have shot eight high and observe carefully. In this way you have a sight picture to carry with you when you proceed to station one for doubles, after dispatching eight low.

Special attention should be given to two high doubles. The lead-off man can move out at right angles to the face of the high house and position himself well back, but in line with station two and the crossing point marker. Watch all four high house targets of the remaining shooters as they shoot their doubles on station one. Once two doubles has been shot, the lead-off shooter has his work cut out for him. Pay special attention to the low house while the rest of the squad shoot their doubles at station two as six low doubles is an oft-missed target and your next shot. After you have shot your doubles at station six watch one or two additional low house targets – this is your next one and you should know exactly what to expect. Only after you have taken your optional can you breathe easier.

20.

Gun safety

Every year there are fatal accidents associated with firearms, both while hunting and in the home. Rarely are these accidents linked with skeet or trap shooters. The reason for this is the safety training and rules laid down by clubs. Skeet and trap shooters handle their guns frequently, and are schooled on the ranges on how to correctly load and unload. Due to experience there is little chance of them accidentally leaving a shot-gun loaded or discharging it at a fellow shooter.

You will find that the following rules closely parallel those known as the "Ten Commandments" of gun safety. I've written them with a special emphasis toward skeet.

a

On arriving at the club when removing your guns from the gun cases, if automatics or pump guns, ensure that the breech is left open. Over and under or side by side guns should be left open. When storing on a gun rack the latter type of gun may be stored closed and automatics and pumps left open.

b

When carrying guns be sure that they are open or broken.

c

Always point the muzzle in a safe direction.

d

Examine the bore of the gun before loading to see if it is clear of obstruction, smaller gauge shells, rags, etc.

e

Do not load until you are on the station and it is your turn to shoot.

f

Unload if a breakdown occurs with the trap equipment. Although there is a safety on a gun do not use it for that purpose.

g

If you have a misfire wait for 10-15 seconds before opening the gun. You may have a hang-fire.

h

Practise good gun safety by loading only one shell at a time when shooting singles.

i

Refrain from shooting at targets that are off the flight-path.

j

Never examine a gun until you have permission from the owner and then check to ensure it is unloaded.

k

If instructing, keep a sharp eye out and be quite firm about all safety practices. Familiarity breeds contempt – don't be careless.

l

If you have a gun breakdown while shooting, keep the muzzle pointed toward the target crossing point and seek help from fellow shooters if you are unable to rectify the situation. Some thoughtless shooters immediately turn round and swing the muzzle around with them. Don't call to attract their attention, merely walk in behind them. "Hold it," "Don't move," is the most appropriate comment at this time.

124 **m**

Due to considerable movement while shooting skeet, there is a chance of inadvertent misdirection of the muzzle of the gun. Assume a correct and satisfactory position for carrying the gun from station to station. The over and under or side by side on the crook of the arm. The auto or pump, on the crook of the arm or the hand at the point of balance, muzzle depressed.

21.

International skeet

No book would be complete without a short discussion on shooting skeet, international style.

The rules for regular skeet, as normally shot in North America, and international skeet, are quite different. I will touch briefly on major differences; should you wish to take up international shooting, obtain a copy of the rules and study them thoroughly.

Of the two types, international skeet is by far the more difficult. There are three basic reasons. Instead of fifty-five yard targets as is standard practice in this country, international targets travel as the rule book says: "to about sixty-five metres or approximately seventy-one yards. The gun position is such that in the 'Ready Position,' you must hold the gun with both hands so that the butt touches the hip and is visible under the elbow." What is more, the gun must be held in this position until the target appears. In addition to this, international targets are not instantaneous but, "shall be thrown within an indefinite period not to exceed three seconds."

For those who started shooting years ago, the gun down and delayed targets will not present too great a handicap. More recent shooters who have always shot with a high gun, that is, with the gun mounted on the shoulder, may take some time getting used to the international position. In normal skeet, targets are usually shot at before they reach the crossing point. In international, they are shot at beyond the crossing point – especially when shooting from stations three, four or five. In fact, in the crossing shots often distance from gun to target is approximately twenty-five yards, instead of the usual twenty-one yards or less. Another variable is the target. The international target is heavier, harder and not as high or thick; in other words, the cross sectional area is less. There is no competition in the smaller gauges; only the twelve gauge. As for shot size, some shooters prefer number eight for international. If you have been shooting a gun barrel bored cylinder on the easier-to-break American targets, you may find on the international target you will require a gun barrel with more choke.

Wearing-apparel for both types of skeet can be much the same but particular attention should be paid to the vest or coat. A normal warm-weather vest with ventilated back may be modified by attaching a flap full length

120. Shooting vest modified for international shooting by attaching a full-length flap

across the front complete with zipper (*refer to photos 120, 121*). A piece of one-inch commercial elastic should be attached to your belt and to the inside of the skeet vest on the right-hand side as in photo (*refer to photo 122*). This arrangement ensures that the shoulder-pad will always present a flat surface with no bunching. You should attach the elastic so as not to have too much tension. Carry the shells in the right pocket if you are a right-handed shooter. If you use a coat, position the elastic on the inside of the coat to get the same results as with the vest. In addition, because greater freedom of movement is required, the under section of the sleeves can be cut open in the area of the armpit to obtain this greater flexibility as in photo 29. Some shooters prefer sweaters, and the elastic from-sweater-to-belt idea can also be applied.

Many shooters have rubber recoil pads on their guns and in this gun down style of shooting, this may present a problem. As the gun-butt is moved from the low gun position to the shoulder, the rubber may catch. Again, if you have a suède shoulder-patch, you may find that the rubber recoil pad may stick; quite often you will not get the butt placed correctly. It is best to shoot with the suède patch on the jacket and a hard composition butt plate which is usually on the gun when delivered from the factory. If your gun has a rubber recoil pad it would pay to cover it with a type of industrial tape. Some shooters often use clear nail-polish on about the top half or two-thirds of the rubber recoil pad to prevent sticking.

121. Modified vest for international shooting with flap zipped up

122. To keep vest from riding up during international shooting, a small piece of elastic with a loop on the end is attached to the inside and the loop is slipped over the belt

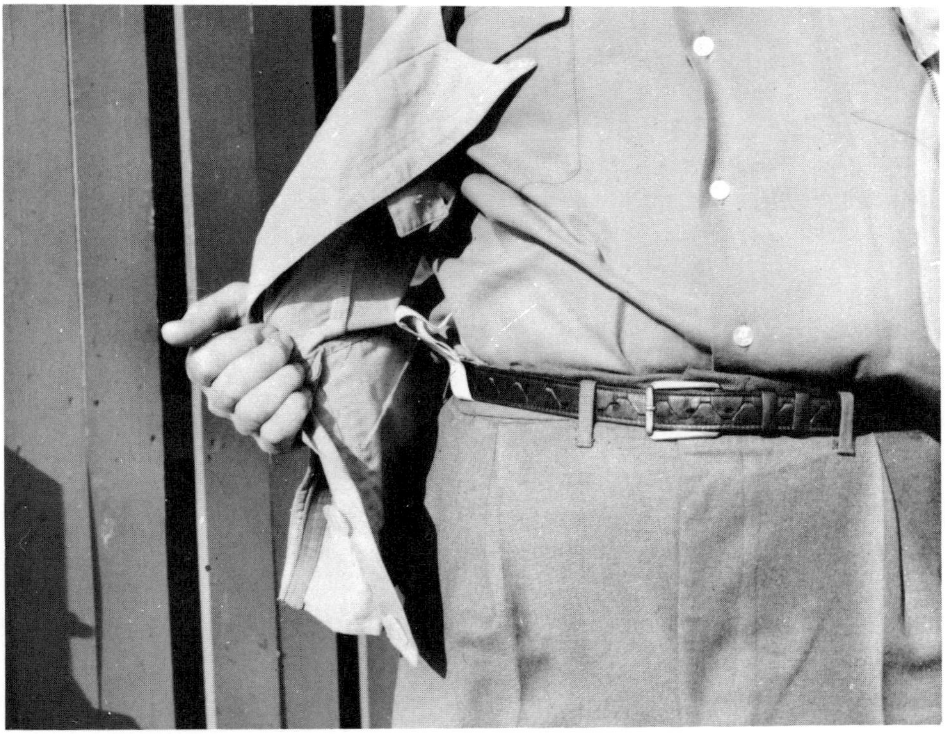

Now come the basic requirements. First and foremost is foot position. Stations one, two, six, seven and eight are the same for international skeet as they are for regular skeet and this also applies to doubles on the same four stations. On stations three, four and five a slightly more exaggerated stance than for regular skeet may be required. Stand a little farther around to the right for the high house target and a little farther around to the left for the low. This again depends on the individual's reaction time, speed of swing and keenness of sight. As the target is much faster it may be better to change the foot position on these crossing shots.

While the rules permit shouldering the gun on stations one and eight, I suggest the second basic requirement is not to dry point on these two stations. The shooter should position the gun in accordance with the rules for international skeet *(refer to photo 123)* and hold the muzzle of the gun so that it is in line with your eye and the crossing point. The muzzle should be such that, when the shooter looks at it and beyond, he is looking at a point fifteen feet above the level of the field. This in fact, is then his dry point. For the dry point at station eight, use the same procedure, and align the muzzle at approximately the same position as for shooting regular skeet.

The third requirement, that of the ready position, is next. As the shooter turns in toward either house, the muzzle of the gun should travel down. As the shooter looks at the muzzle and beyond, it will be on, or slightly below the intended flight-path of the target. The big question here is: How far back should one go? This again will depend on the area background and the reaction-time of the individual. On all crossing shots, do not swing back quite as far as you would on normal skeet.

Shooters who are not accustomed to shooting with the gun down will have trouble at first in bringing the butt up to the shoulder without causing the muzzle to dip down in a see-saw motion. Keep the muzzle going along the flight-path of the target and bring the butt up so that by the time you have caught up with, or established, a lead, the gun will be properly placed on your shoulder. Quite often shooters in the early stages jerk the butt up to the shoulder and then start out after the target, which by this time is over

123. Ready position for internationa skeet

 the crossing point. This motion doesn't only look bad – it is bad. The flow should be smooth. You may find that while the left hand should normally take the same position as for regular skeet it may be advantageous to hold it a trifle farther back along the fore-end. This may assist in giving a little faster swing to the gun. And speaking of guns, if you have been in the habit of shooting with a high stock you may find that you will require one with a little more drop. When shooting international skeet you cannot wrap yourself around the gun as you do for normal skeet. By and large you tend to stand up a little straighter while in the ready position and therefore a slightly lower stock may be better.

Now we come to the question of lead or forward allowance. The international target is approximately a third faster than the target for regular skeet and the distance from the gun muzzle to the target when the shot is fired is also greater. With all the variables associated with this style of skeet shooting it is impossible to set down a table of leads similar to that for regular skeet. It is enough to say here that leads have to be increased.

A very important aspect of competitive shooting is the East-West time zone change. Doctors say that in an East-West time zone change of four to six hours it takes a person at least a week to get their system back to normal. This means that if a shooter is involved in a competition shortly after arriving at his destination he will not be in the best physical shape. While international skeet is now on the increase in America, the term "international" has in the past implied shooting abroad. Changes in food and drink can also adversely affect shooting.

Normally at home, you shoot with four friends you've known for years. The time taken to shoot one hundred targets is about one and a half hours. You select your time for shooting and your position on the squad. There are no last-minute changes in rules due to misinterpretations by some countries and "Dead" and "Lost" are the calls from the referee for scoring the targets.

International shooting is entirely different. You find yourself in a strange country, eating strange meals with hundreds of athletes of all types. The tension is greater and the surroundings make it even more so. You shoot alone without your friends as you are squadded by chance and you have no choice of position in the five-man squad. The time to shoot one hundred targets will depend on the number of fields available and the number of contestants. In Oslo in 1961 it took twelve hours to complete the hundred targets on the first day and the same on the second. There were some eighty contestants and only two fields, and with delayed targets it added to the time required to complete a single round. You are employed as assistant to the referee at least once or twice during the hundred target event. A horn is used to signify a lost target and you can be assured, if you've never heard one at your club, that it's most disconcerting. You can be fined targets for being late or lose targets if both of your feet are not within the marked confines of the three-by-three square of the station. There are also delays and conferences which arise due to some shooter or behind-the-scene associate calling "Protesto." These, along with gun down, delayed timing and faster targets, add to the intrigue of international shooting.

22.

Flinching

No greater misfortune can befall a skeet shooter than to be attacked with a case of what is known as "flinching." All of a sudden it starts; it seems to be a temporary paralysis of certain nerves. There are two types of flinching, the psychological type where the conscious mind says pull the trigger and shoot while the subconscious says no. There is a never-ending battle and the shooter rides his targets clear across the field and finally pulls the trigger. At this precise instant he leans forward into the gun in anticipation of the recoil. His forward movement induces an additional flinch which invariably results in undershooting. Physical flinching, the second type, may go hand-in-hand with the former or psychological flinch or it may appear by itself. Physical flinching occurs more often than psychological and both can be associated with the state of one's health.

Flinching does not attack new shooters but usually takes its toll of the more mature ones. The new shooter is kept busy just trying to break targets and as such has no inhibitions. The older shooter has had many ups and downs, has read everything on skeet and shooting generally, has changed his guns, adjusted his stocks and has done everything possible to improve his scores. His shooting may have slowed down slightly and he has started riding his targets. Like a thief in the night, all of a sudden flinching occurs without warning. The more you worry about it, the more conscious you become – the more the shooters talk about it, the harder it is to correct.

Much has been written on how to correct flinching. If a shooter has suddenly become afflicted because of a change in timing generally, a switch back to a faster timing is all that is required. So often one's health may be the cause, or overshooting, or it may be due to shooting a gun which does not fit properly. In these cases the best thing to do is to stop shooting and get your nerves and physical system back in shape. During this time carry out a thorough investigation to find out what is causing the problem. Examine the equipment that you are using: Does the gun fit properly; is the stock too high; is there too much pitch down? Are you using the correct recoil pad or is it too soft? What about the recoil pad in your jacket – sometimes soft rubber permits the gun to bounce? If you have recently changed from winter to summer shooting, possibly the shoulder-pad is too thin after shooting all winter with heavier clothing. Have you shortened the grip on the left hand? Perhaps you were holding farther out, thus absorbing considerable recoil in the left hand. You may also have shortened your grip, allowing the gun to pound your shoulder more. Have you been shooting heavy load shells at some other sport, *i.e.* crow hunting, fox hunting, etc.? Perhaps you have injured your elbow or your back in some extracurricular activity.

After a rest, start out with a smaller gun such as the 410 and if everything is normal, once again work up slowly to the larger gauges. Sometimes switching to a recoil reducing device such as a hydro-coil or a compensator helps, but beware of the latter device as there will be an increase in muzzle blast. If you have been shooting with a ventilated compensator this may have resulted in the flinch originally, due to excessive muzzle blast.

Last but not least is the question of trigger pull. Flinching can be precipitated by a heavy pull and anything in excess of five pounds should be considered heavy. It is recommended that a trigger should be set for a four-pound pull. It should be a crisp pull with no long take-up. If you have gone to a new gun or if your trigger or sear has become worn, both excess play or take-up and heavy trigger pull may be your downfall.

23.

Targets

For night shootings, targets should be all white. In daylight hours, targets should be all black – providing you have no background problems. For skeet ranges, visibility and background are most important and most skeet fields are set out with this in mind. Lots of skyline free of hills and trees makes the best background. When targets are viewed from various stations you see only the silhouette. Why sacrifice this outline with a yellow or white crown? Both types blend into the background and the target appearance is then naturally reduced in size. If all black, the targets will look larger and be easily seen, while the scores will be higher. Also, if the paint comes off on the rubber rail, positive spin will not be given to the target at all times. In winter, the paint forms a glassy surface in the frozen rubber rail and thrown targets will vary a considerable amount in the horizontal. Targets free of paint will be more easily seen and cause less trouble in both summer and winter.

Most of our backgrounds are yellowish-green or whitish with the fleecy clouds. Why is it that on an overcast or foggy day you shoot so much better than with white and grey clouds and clear blue sky all mixed up? The reason is that you see the full silhouette on the overcast or foggy days. If the targets are completely black they look as big as footballs. Try all-black targets on your fields – you'll see it's an advantage.

There are of course some locations, such as Reno and the hills of California, where you need a yellow- or white-crowned target. But in the majority of cases the all-black target is an advantage in more ways than one. The white or yellow crown is an off-shoot from trap shooting, where usually there's a period of time when the target blends with the background from the top of the house to the horizon. This is where the trap shooter wants a dissimilar colouration of his target. Then again he sees a partial plan view, an ellipse, and in this case colouration of the dome is an advantage except in winter. Where there are areas of bad visibility and yellow or white targets do not seem to show up well, clubs should experiment with other colours to achieve the ultimate in target contrast.

APPENDIX

Lead or Forward Allowance

In shooting skeet the amount of lead required for any shot will depend, within certain limits, on the individual's reaction-time, the fit of the gun stock, the speed and type of swing and the keenness of sight of the particular shooter. For this reason no one standard can be set up which will apply to all persons. The statements and figures which follow concerning each station and shot are not to be taken as inflexible rules. They closely approximate what will apply to the average shooter and form a working basis from which any shooter by careful practice and study of the results, may determine his own personal equation for each shot.

The approximate leads for all stations are based upon the premise that all targets will be broken at or about the crossing point except those shot from station eight. Some shooters allow incoming targets on stations one and two, and six and seven to ride in before shooting at them. Under these conditions minor changes in lead to that recommended will be required.

STATION	HIGH HOUSE TARGET	LOW HOUSE TARGET
1	Directly at or slightly under	One foot
2	One foot	One and one-half feet
3	Two feet	Three and one-half feet
4	Two and one-half to three feet	Two and one-half to three feet
5	Three and one-half feet	Two feet
6	One and one-half feet	One foot
7	One foot	Cover the target
8	Cover the target (continue swing)	Cover the target (continue swing)

Additional Comments

STATION ONE HIGH

A fast shooter in going down with the target usually shoots directly at it, while a slow shooter (slow reflexes) will want to shoot just under the target by not more than four to six inches.

STATION ONE LOW

Most shooters shoot at this target after it has passed the crossing point at the centre of the field, and many let it ride right in. Shoot just in front of the target and use about a foot lead. Again, a shooter with fast reflexes, coming from behind the target with the muzzle of his gun, will fire just after he passes the target.

STATION TWO HIGH

A shooter with fast reflexes and using the swinging-through method will fire the instant the muzzle passes the target. The slower shooter should use one foot lead or just daylight the target.

STATION TWO LOW

Daylight the target by using a lead of one and one-half feet. The shooter using the swinging-through method will fire just as the muzzle of his gun passes the target.

STATION THREE HIGH

The lead on this station is approximately two feet; if you must err, err on the long side. You can overlead by a third of the required lead and still break the target. This holds good on all stations. Conversely, if you are late it will result in a lost target.

STATION THREE LOW

This is the target, along with five high, which requires the maximum lead. You should use a lead of three and one-half feet.

STATION FOUR HIGH AND LOW

A lead of two and a half to three feet is required on either of these targets.

STATION FIVE HIGH

Use maximum lead on this target, three and one-half feet. Target flight-path is at right angles, the same as three low.

STATION FIVE LOW

Lead this target by two feet. Keep the gun swinging.

STATION SIX HIGH

Lead this target by one and a half feet and shoot it right after it passes the centre stake. Many good shots ride this target well in.

STATION SIX LOW

Again the fast shooter will fire just as the muzzle of the gun passes the target. The slower shooter should daylight the target or lead it by approximately one foot.

STATION SEVEN HIGH

Shoot this target just as you did on one low. Use about one foot lead, if using the swing-through method fire just after the muzzle has passed the target.

STATION SEVEN LOW

This is a rising target, be very careful and don't hurry. Look at it well before shooting. Most misses occur by undershooting. While other members are shooting at this target observe the general flight-path and angle of elevation. When it is your turn, put the gun on this alignment, cover the target and fire.

STATION EIGHT HIGH

Cover this target and fire, be sure to continue the swing.

STATION EIGHT LOW

Cover this target and fire, be sure to continue the swing.

NOTE

The major fault on both high and low house targets on station eight is that the shooter likes to see the target from the time it leaves the trap until he fires at it. If he can still see it when he fires, chances are that he will miss by being late.

DOUBLES

The leads required when shooting doubles are the same as for singles. A shooter usually shoots with a faster reaction-time and as a result he may hit his doubles better than his singles. The swinging-through method is the best, although for the second target at stations two and six, the pointing-out method may be used but is not recommended.

Acknowledgement

There are between sixty and eighty persons, mostly shooters, who in one way or another, have contributed to my successful career in the competitive shooting sport of skeet. It is impossible to name them all here but I would just like to mention a few.

First and foremost is my wife Johanna or "Joey" for short. She has put up with my whims and idiosyncrasies and thereby permitted me to spend hours on the practice range or at some far-away tournament. To be able to do this, if married, you need a very understanding wife.

My close companion for many years and two-man team partner "Cup" McCreery of Ottawa, Ontario, helped in so many ways and I am deeply indebted. The late Dunc Morrison of Canadian Industries Limited enabled me to enter and win my first competition, the Halifax Bicentennial in 1949. Tom Ross, Dick Tobin and Frank McCormick, also of C-I-L, encouraged me to set goals; new ammunition for international shooting and the now famous 410 gauge flat crimp target load were among their contributions. I attribute a good measure of credit to C-I-L's top quality ammunition of all types, which helped me set a number of world records. Bill White of the Montreal *Gazette* and Stuart Molson, past president of the N.S.S.A., also provided me with competitive opportunities and encouragement for which I am grateful.

There are others such as Ed Scherer and Bob Bogie who by what they said and suggested helped me to become a better shot. To Curt Clair and Jack Millhofer of Remington and their helpful gunsmith Neil Simmons; to Fred Bates and Cliff Doughman of Winchester; to Roland Pacquin, Peter Wilson and Harry Heiter of Browning and Ernie Simmons of Simmons Gun Specialties – many thanks for your courtesies and patience.

I must not neglect to mention Cpl. Al Santorelli (ret.) and C. W. D. "Ed" Edmiston of the U.S.A.F. when they were stationed in Goose Bay, Labrador in 1946-49. Johnnie Beggs S/L (ret.) of the R.C.A.F., the armament personnel of Station Rockcliffe and the R.C.A.F. generally have earned my gratitude, particularly A/M Hugh Campbell (ret.) and A/C the late Joe Stephenson. I would like to especially mention A/C Pete Gilchrist (ret.), with whom I shot constantly and who gave me immeasurable help.

To these and to many others all I can say is *thank you.*

BARNEY HARTMAN S/L (RET.)
SHOOTING PROMOTION REPRESENTATIVE OF C-I-L
1967

World Records and Achievements – B. C. Hartman

1966 *All-time, all divisions, high all-around rating of .9875 set on 2500 targets with 12, 20, 28 and 410 gauge guns (World Record).

*All-time, all divisions, 410 gauge rating of .9875 (World Record).

100/100 – 410 gauge Nationals at Rochester, N.Y. (World Professional Record). Only Canadian and Professional to post perfect score.

1963 *All-time, all divisions, 12 gauge rating of .9982 (World Record).

1962 200/200 – 1962 Nationals at Montreal, P.Q. with four guns, Champion of Champions event – (World Record).

1960 200/200 – two man team, McCreery-Hartman (Record shared by many).

1959 200/200 – two man team, McCreery-Hartman (Record shared by many).

348/350 – 100/100 (12 gauge), 100/100 (20 gauge), 99/100 (28 gauge), 49/50 (410 gauge) – (World Record).

1958 *All-time, all divisions, 20 gauge rating of .9950 (World Record).

1956 442/450 – 246/250 (12 gauge), 99/100 (20 gauge), 97/100 (410 gauge) – (World Record). Shared by Ben di iorio – *same shoot.*

1954 200/200 –100/100 (12 gauge), 50/50 (20 gauge), 50/50 (410 gauge) – (World Record). Shared by Dave Crosby, who shot his in 1959.

*NOTE:
Holds the world high average record for all-time and all divisions in four of the five categories. This covers the best performance in any one year. See lifetime average performance.

Lifetime Averages

(Based on at least fifteen years in registered shooting)

410 gauge	93.65%	—	2nd in World
28 gauge	97.26%	—	2nd or 3rd in World
* 20 gauge	98.38%	—	1st in World
* 12 gauge	99.21%	—	1st in World
*All-around (4 guns)	97.73%	—	1st in World

*World Records

World Championship Performance
(attended from 1959 to 1966 inc.)

410 gauge	—	764 x 800	= 95.50%
28 gauge	—	787 x 800	= 98.375%
* 20 gauge	—	796 x 800	= 99.50%
* 12 gauge	—	1995 x 2000	= 99.75% (inc. run of 1322 straight)

*All-around (4 guns) = 98.72%

*probably World Records (no known records)

The above shows conclusively that Hartman has been the best all-around performer in the history of the sport!

All-Star Teams

By the time the 1967 team was named (on 1966 performance) he had been on ten consecutive All-American Skeet teams – 1958-67 – six amateur and four professional. He will have been captain of two amateur (1960 and 1963) and two pro teams (1964 and 1967). In the four years eligible, he was four times captain of the All-Star Canadian Skeet Team – 1960-63.

The Canadian Skeet Championships were inaugurated in 1956 and he won the title for seven consecutive years – 1956-62. The listing of individual titles won means little compared to the story told by the above records and performances.

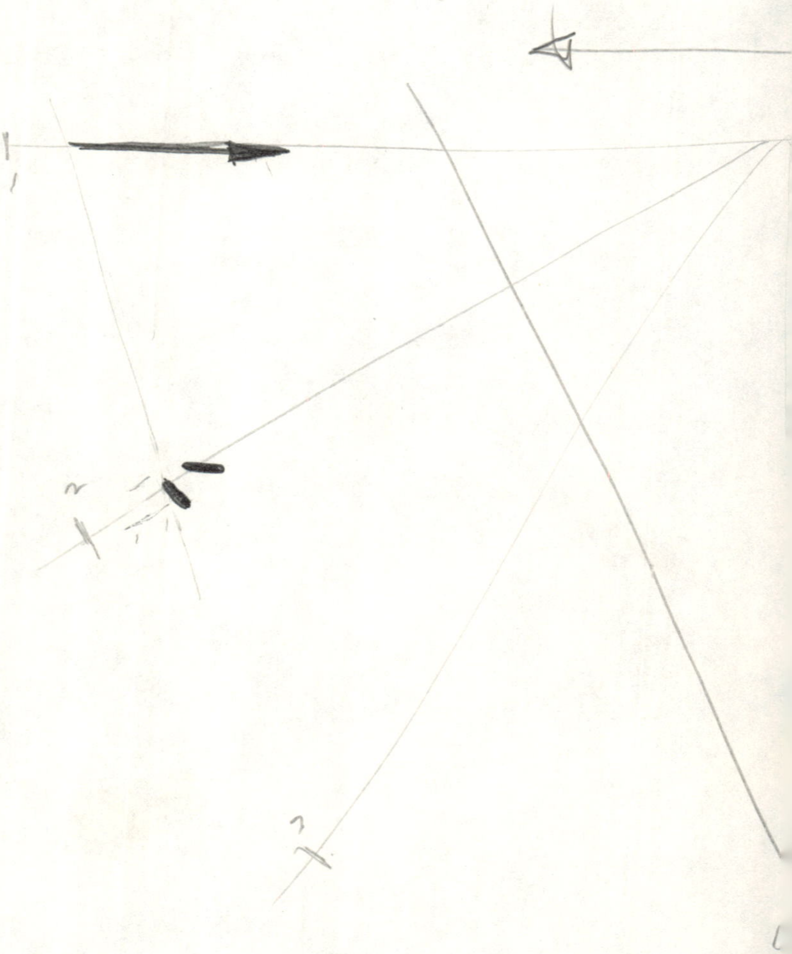

4
6
5